ROBERT SHULTS

THE SUPERLATIVE LIGHT

Daylight

Cofounders: Taj Forer and Michael Itkoff
Designer: Ursula Damm
Copy editor: Laura Fredrickson

ISBN 978-0-9897981-2-9

Printed in China

Daylight Books

E-mail: info@daylightbooks.org

Web: www.daylightbooks.org

Introduction

By Todd Ditmire

I have worked with light my entire career. As an experimental physicist who works with ultra-high-power lasers and studies how intense light interacts with matter, I have spent my professional life manipulating laser light, aligning it through a complex maze of optics, amplifying it in laser crystals, directing it through tubes, compressing its pulses down to unimaginably short times, focusing it down to tiny spots, and propagating it into high-temperature gases and plasmas. These research activities take place on a stage that is the laboratory, a stage filled with fascinating characters: high-technology electronics, large vacuum tanks, high-voltage cables, and the bright, talented scientists who design and use this high-tech menagerie. These are the players in the daily drama of modern experimental physics research.

But it was only with my first viewing of Robert Shults's photos presented in this book that I came to realize that light—the mercurial electromagnetic waves which are the instrument of my research—is itself a very real character in the cast that passes in and out of the laboratory. It was an exciting realization.

That there is an aesthetic beauty in the laboratories in which we physicists work is no new revelation to me. Since my early days as an undergraduate in physics, I have been attracted to the exotic forms of experimental physics apparatuses. I have always been particularly fond of the beautiful complexity seen in a forest of glass elements and mounts on an optical table or in the shiny, futuristic forms of vacuum chambers and high-pressure tubing. I often use images of these laboratory instruments as desktop pictures on my computer or as background photos on my iPhone. But Shults's images have shown me the beauty of our labs in a new way, revealing a dark and mysterious aspect to the alluring images of high technology that serve as the backdrop for the research work I and my colleagues perform.

Shults's vision of the lab is fascinating. It presents what is otherwise a sterile setting in an atmospheric manner that I could never have expected. The location for these photos is the Texas Petawatt Laser, a research facility whose construction began in 2004 as part of the Center for High Energy Density Science at the University of Texas at Austin. The goal was to build the world's highest-power laser and make it available to students and scientists in Texas and around the world to study exotic and extreme states of high-temperature matter. The laser works, in essence, just the same as the modest laser in a laser pointer or compact disc player: it amplifies light. At the Texas Petawatt lab, light pulses whose initial energy is no more than the kinetic energy of a mere mosquito are amplified by a factor of a trillion. Then these pulses of light are compressed in time to flashes that last only an ephemeral instant.

The result is a device that yields parameters akin to what one might expect to find only in a science fiction novel. The Texas Petawatt laser, when it fires (an event that occurs perhaps seven or eight times in a working day), produces pulses of light with a peak power of 1,000 trillion watts, a power 2,000 times that of all the power plants in the United States. This almost unfathomable power, produced for only a fleeting instant, is achieved by clever manipulation of the electromagnetic waves that make up the light pulse, compressing the energy of the laser burst (an energy comparable to the energy in a candy bar or a cup of coffee) to an unimaginably short period of time. The pulses emitted at the end of the Texas Petawatt laser chain are compressed to a tenth of a trillionth of a second, a duration that is the same fraction of one minute that a minute is a fraction of the age of the universe. These are essentially the fastest man-made events.

This flash of high power can then be focused, concentrated to a spot about one-tenth the width of a human hair. When those trillions of watts of power are concentrated to such a pinhead spot, the light is comparable in brightness to the light emitted by some of the most exotic events in the modern universe, including gamma-ray bursts, the enigmatic explosions observed in faraway galaxies thought to be driven by black holes. When focused onto a target (or, at times, into a gas), the Texas Petawatt laser pulse can create temperatures that exceed, by thousands of times, the temperatures found at the center of our sun. Such a laser lets us study matter that can be found naturally only in some of the most remote and inhospitable places in the universe. And this we can do right in the heart of our lab, affording us an exciting chance to study matter that is like nothing found on earth.

Over the years of my career, I have never lost my sense of excitement and awe inspired by the laboratory machines and the experiments they make possible. Shults's photos reveal to me a new way to look at the lab and the people who work in it. His vision of the facility beautifully captures the sense of mystery and wonder that I have always felt about this fascinating stage where I get to perform, a stage that I now see is populated by one more fascinating character.

Todd Ditmire earned his bachelor's degree from Harvard University in physics and art history and received his PhD in physics from the University of California at Davis. He is currently a professor of physics at the University of Texas at Austin, where he is director of the Center for High Energy Density Science and the Texas Petawatt Project. He was previously a staff scientist and project leader at Lawrence Livermore National Laboratory.

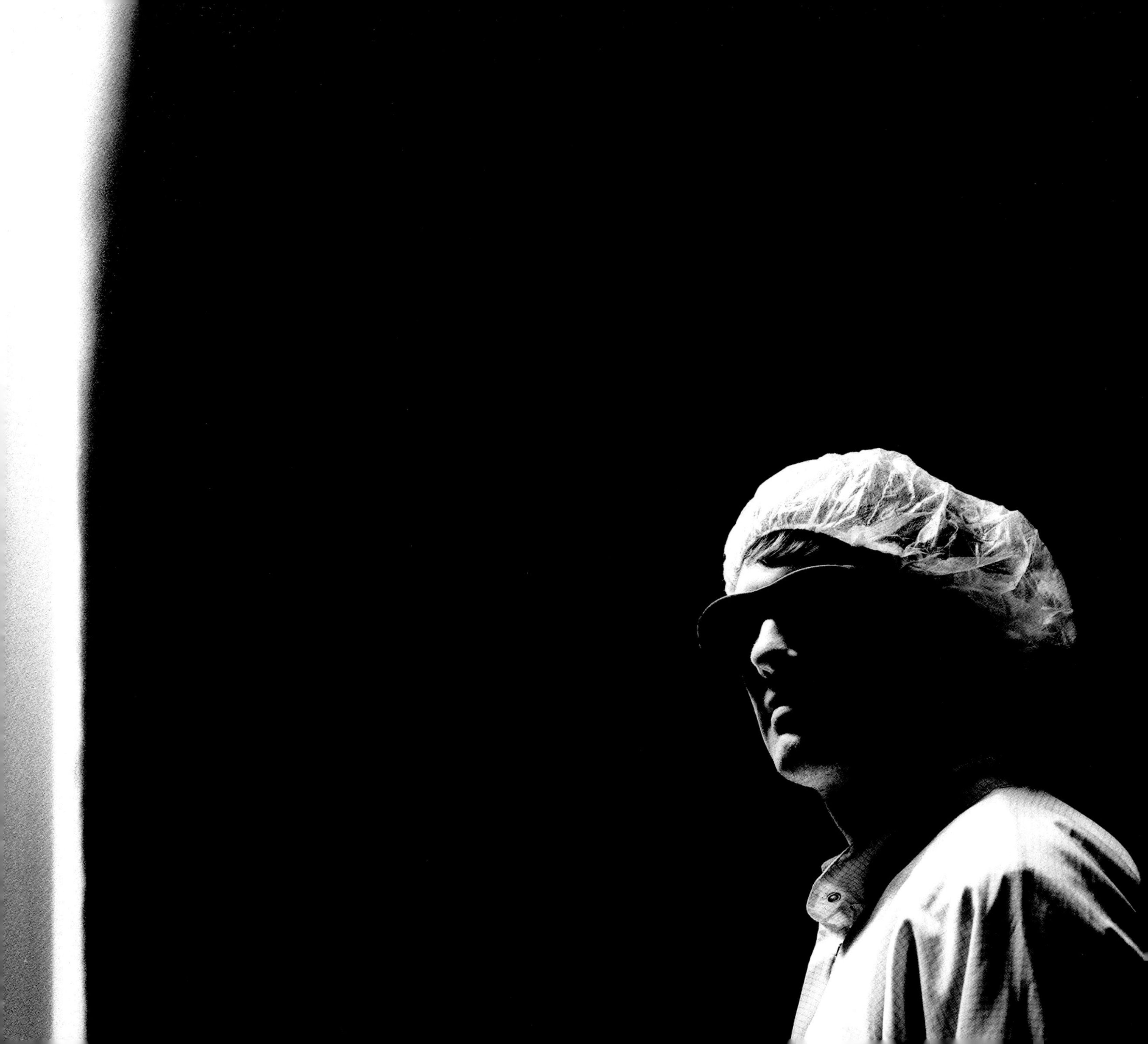

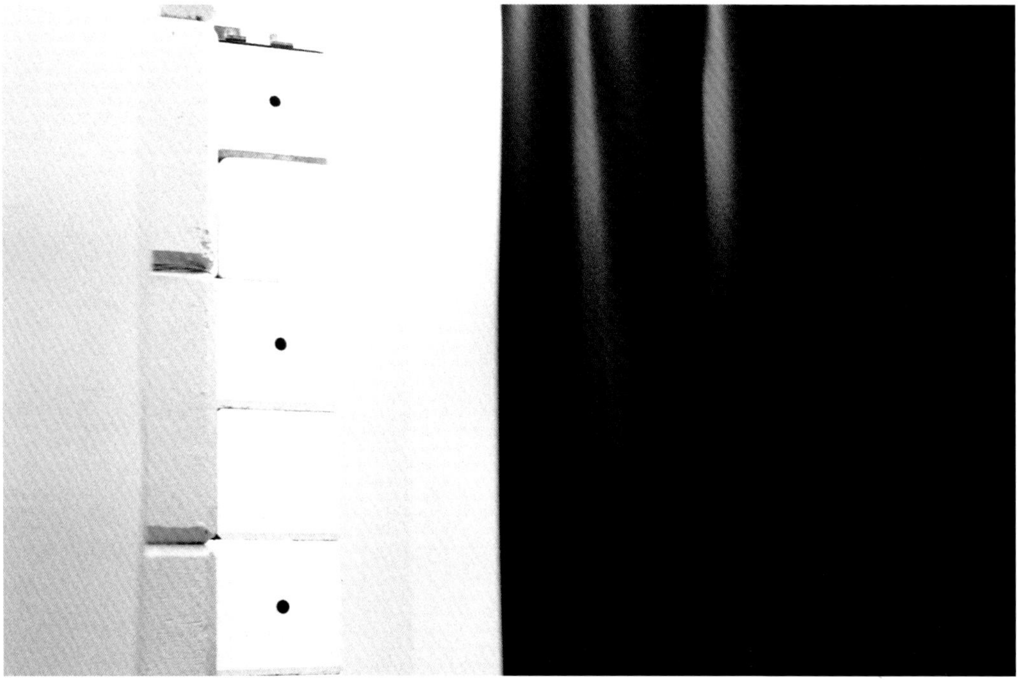

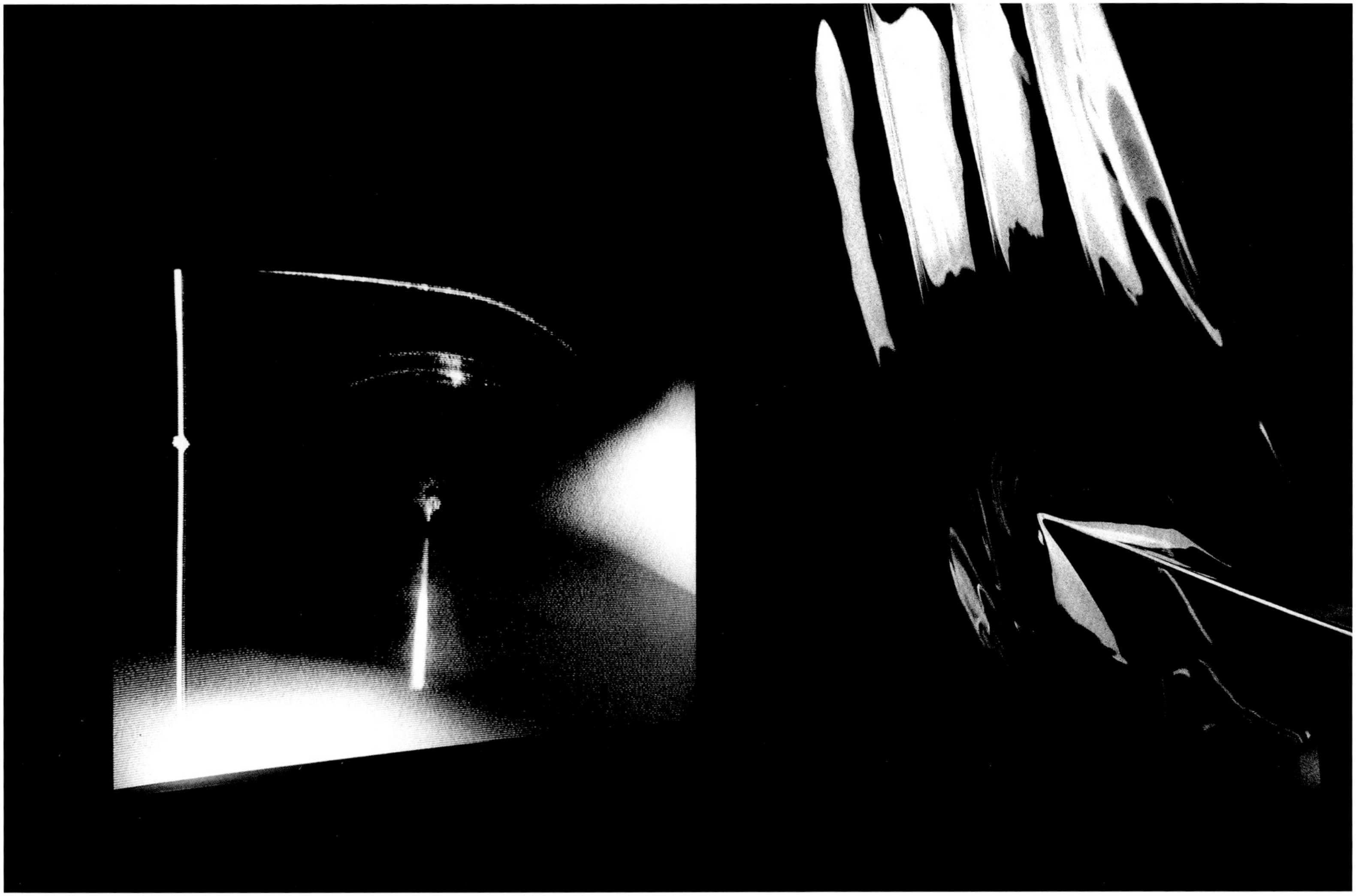

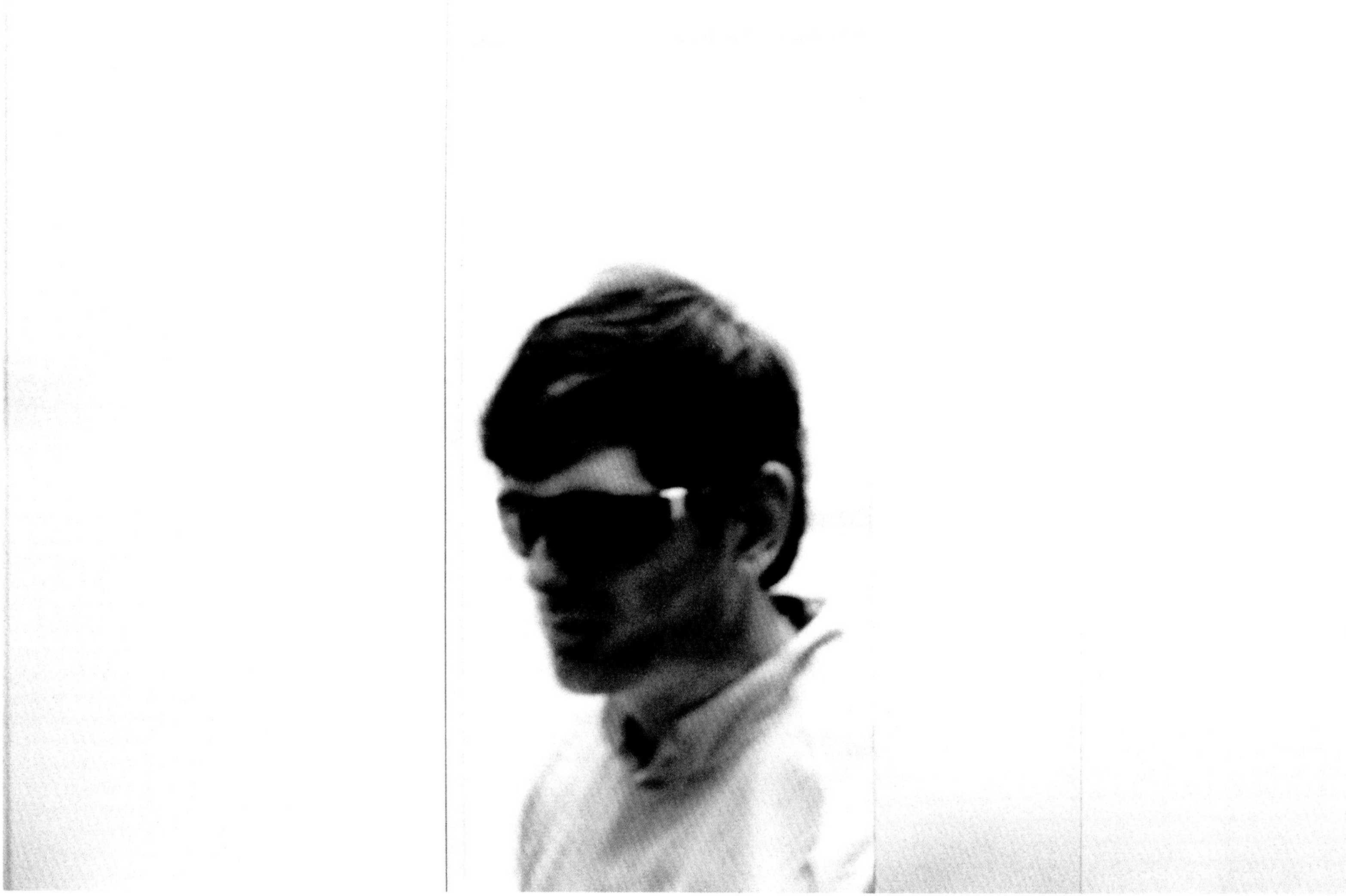

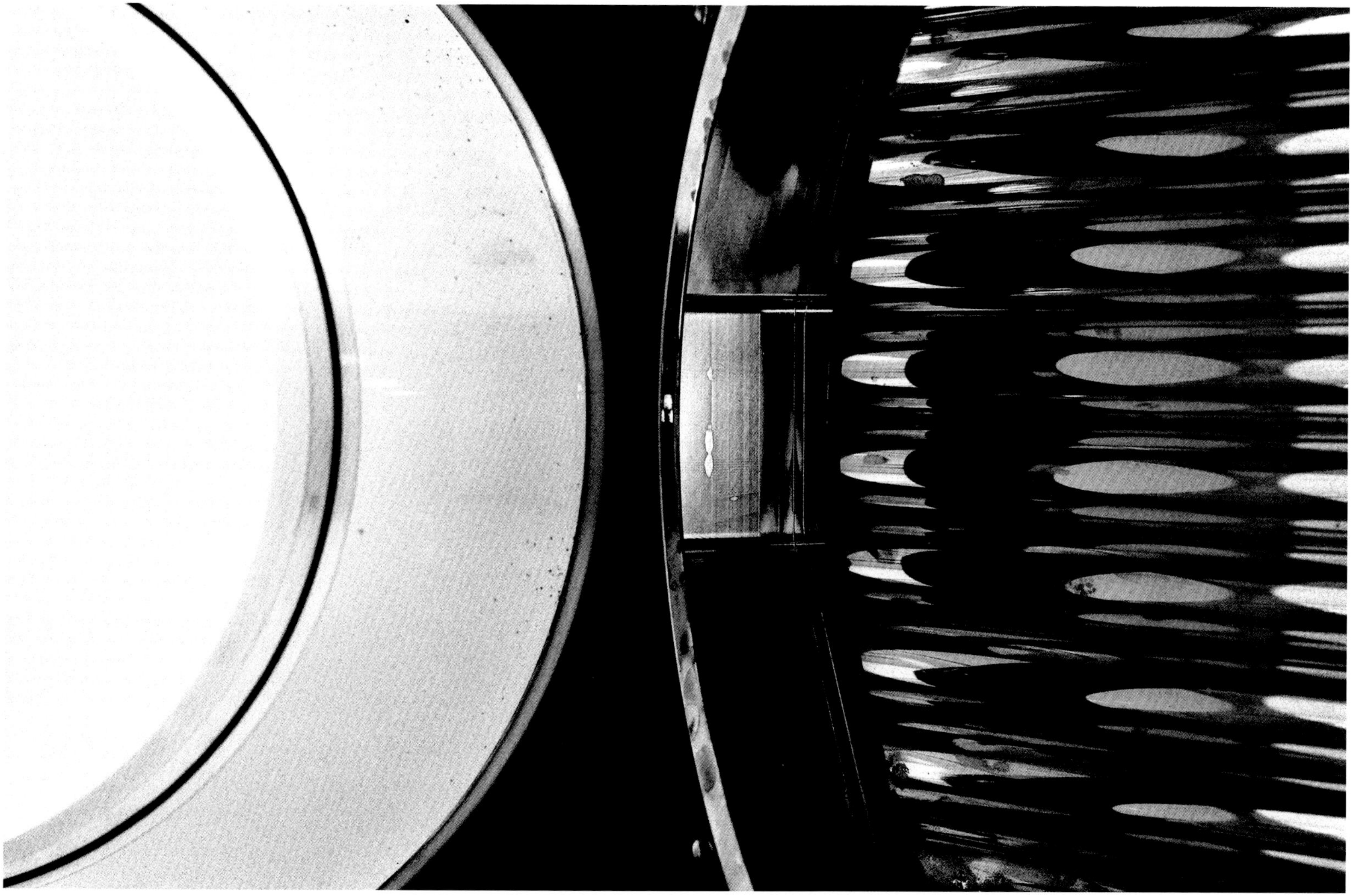

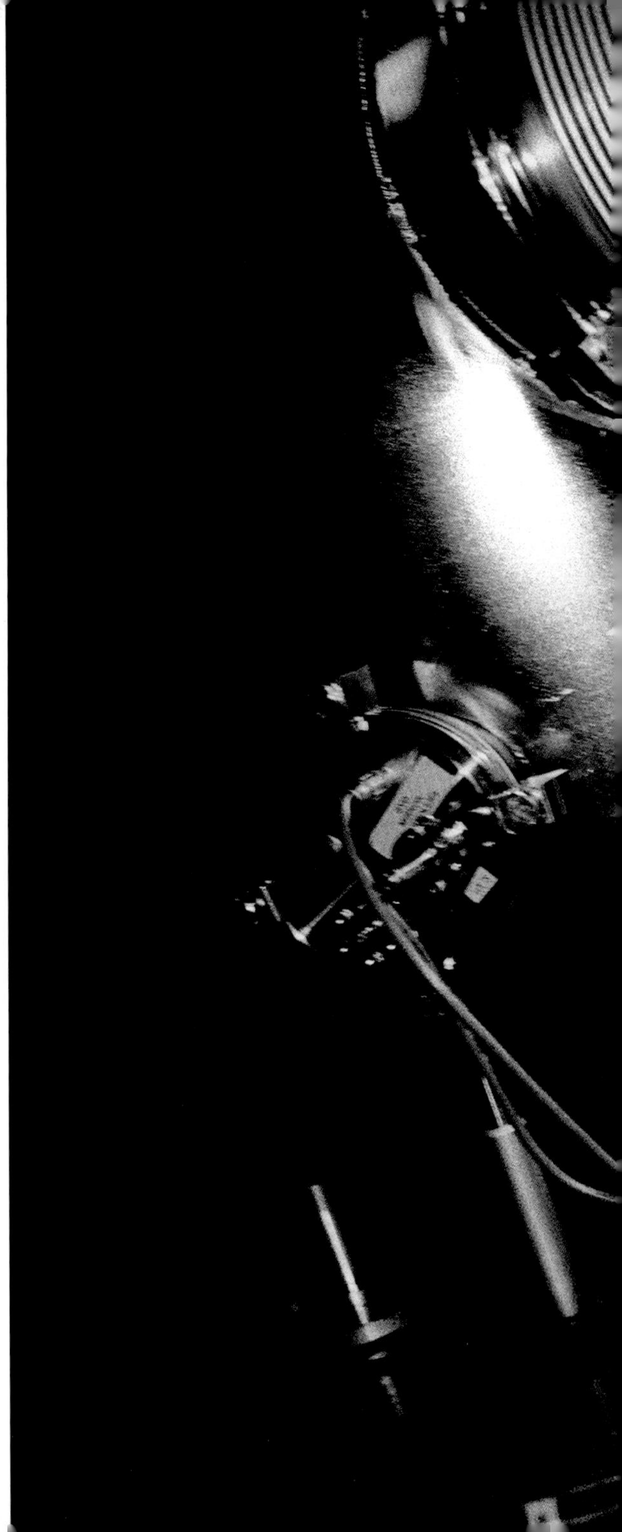

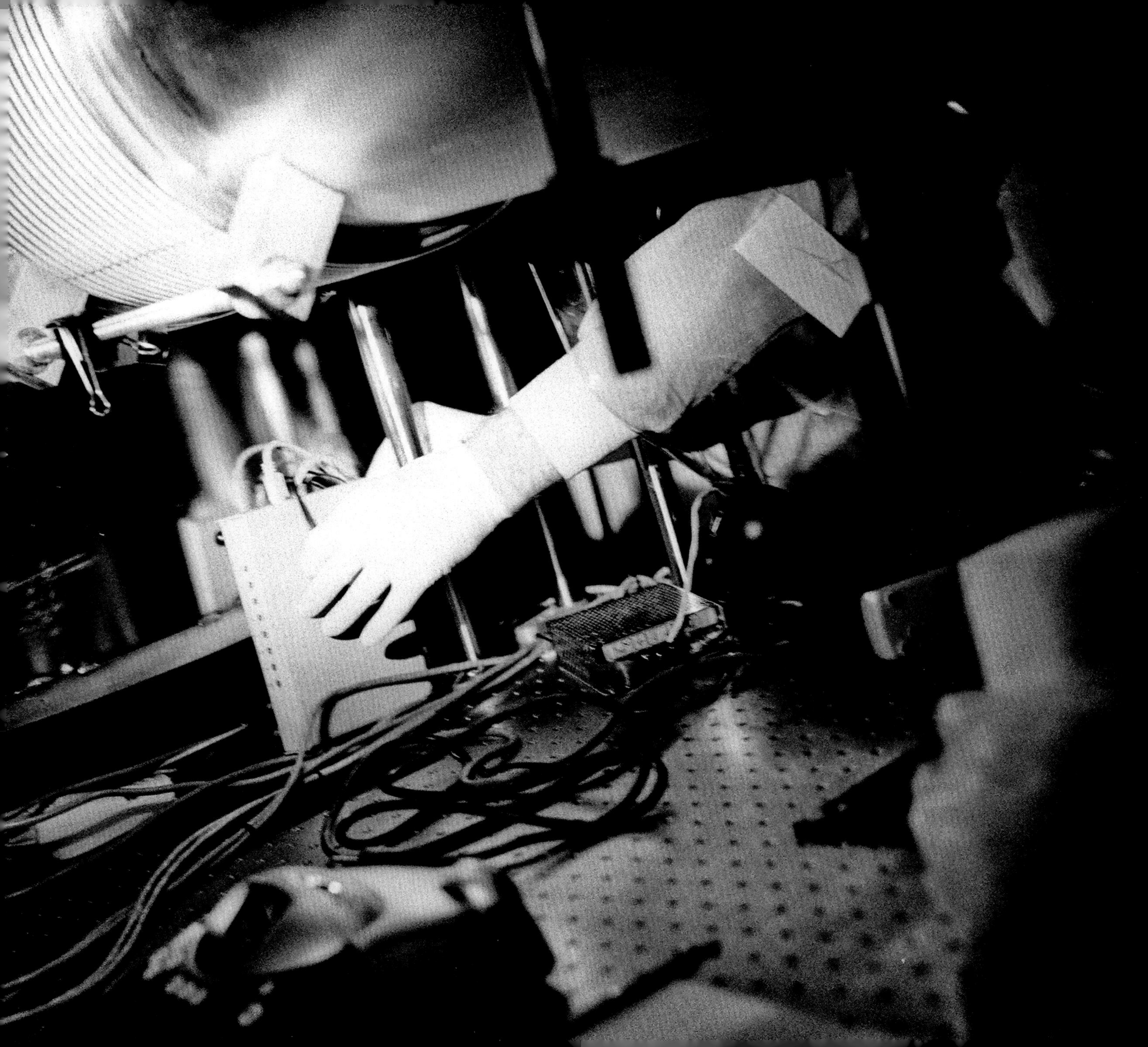

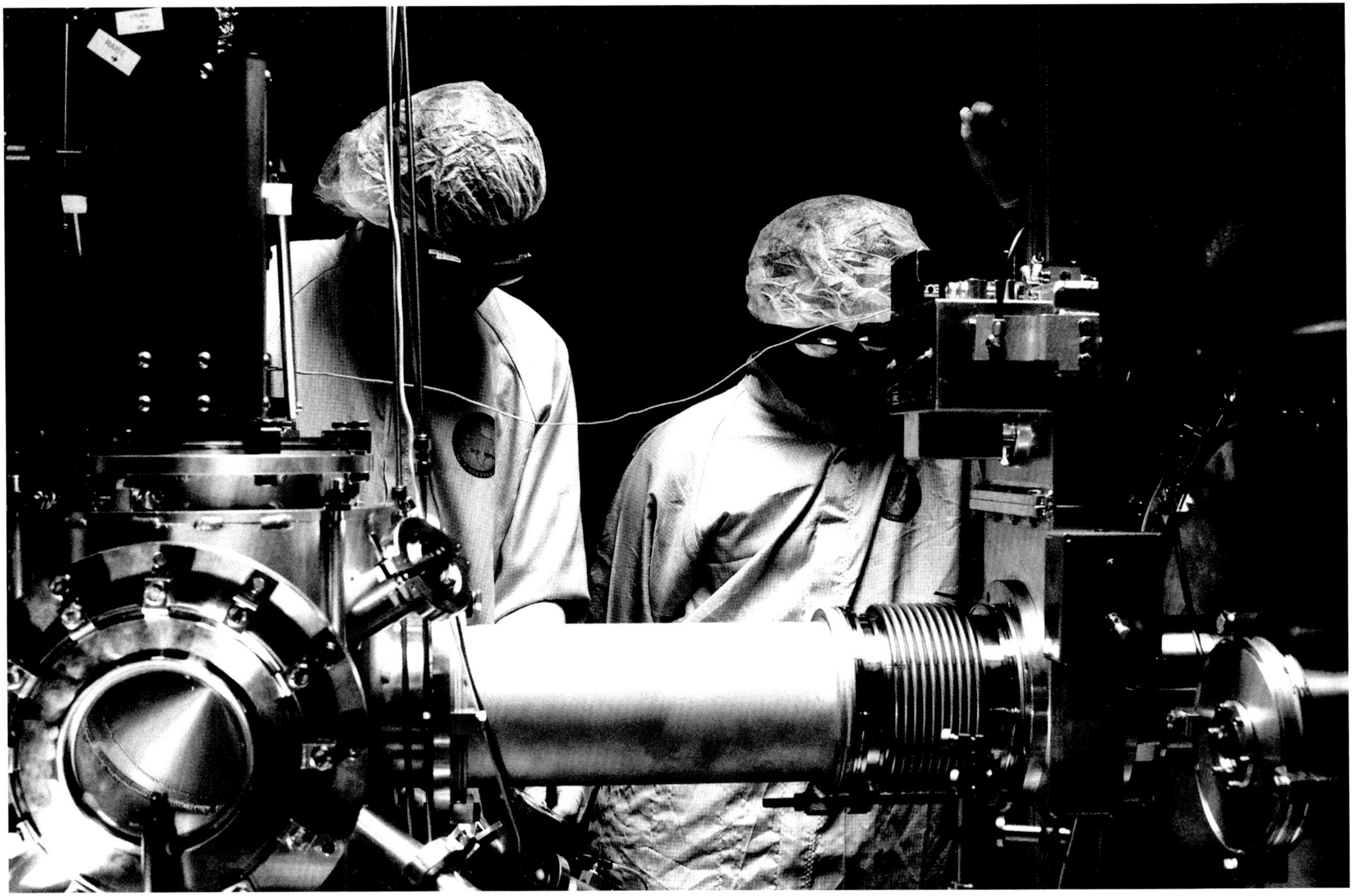

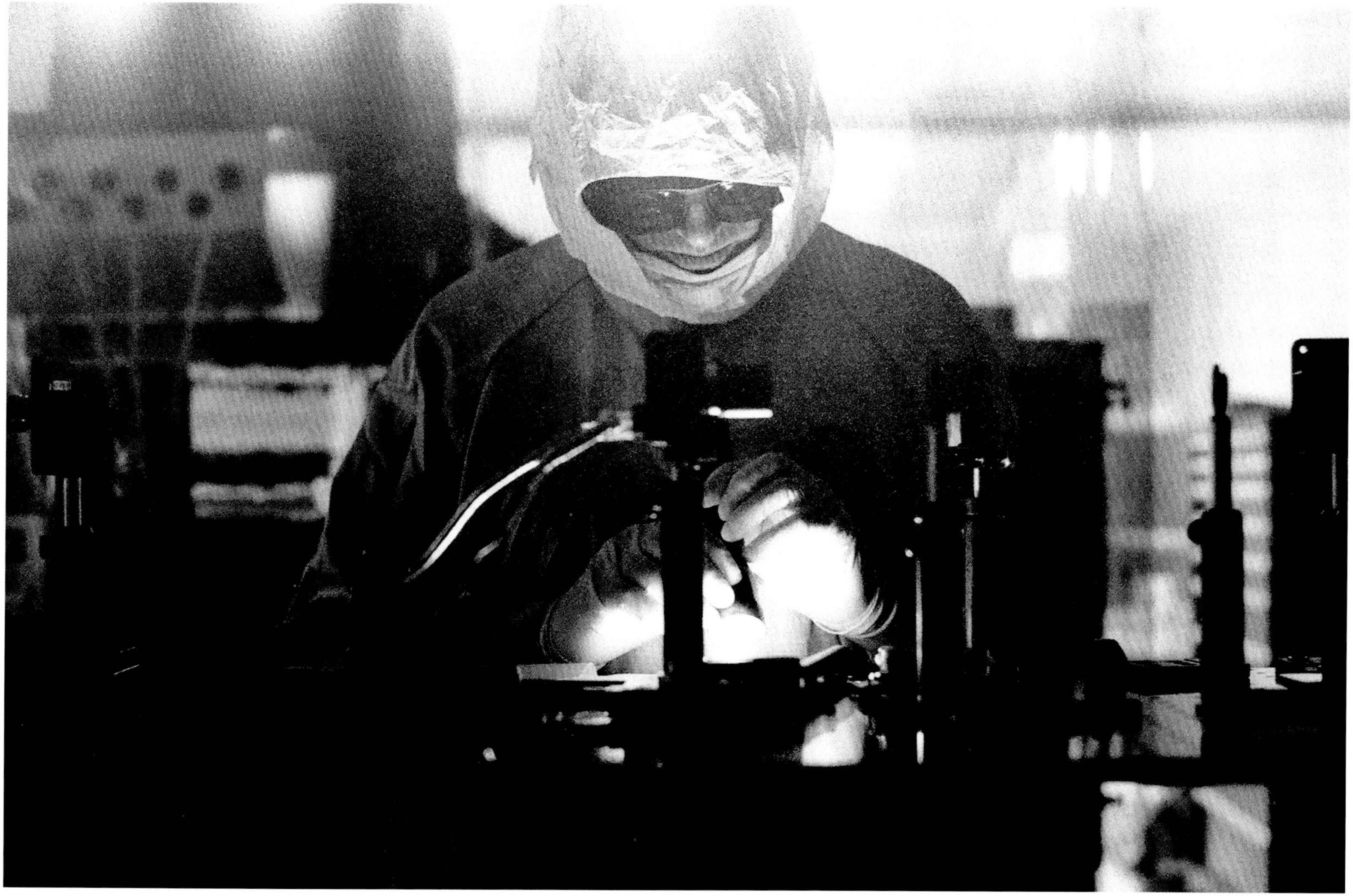

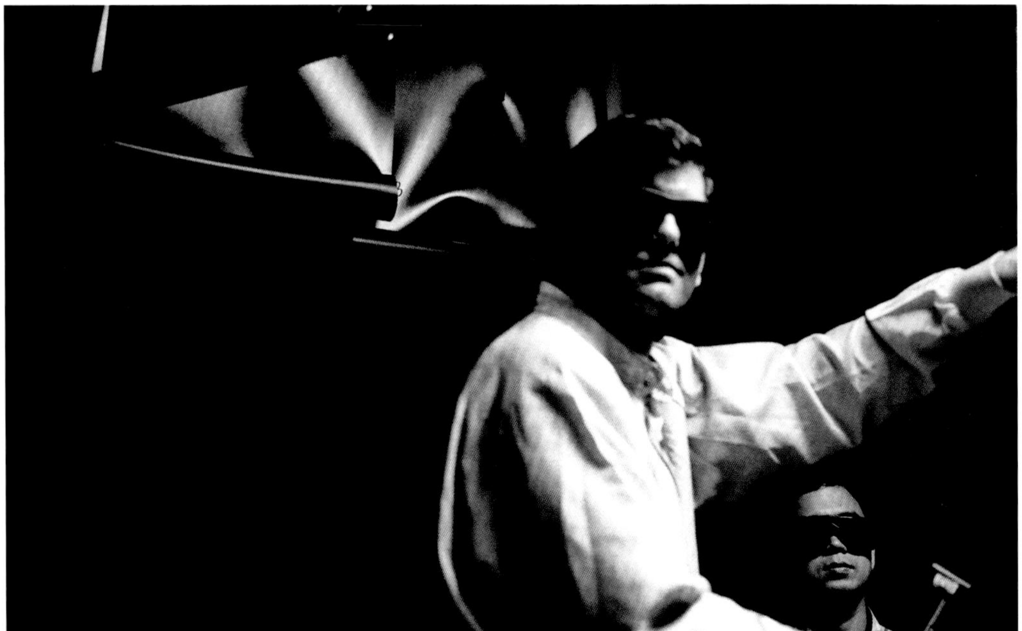

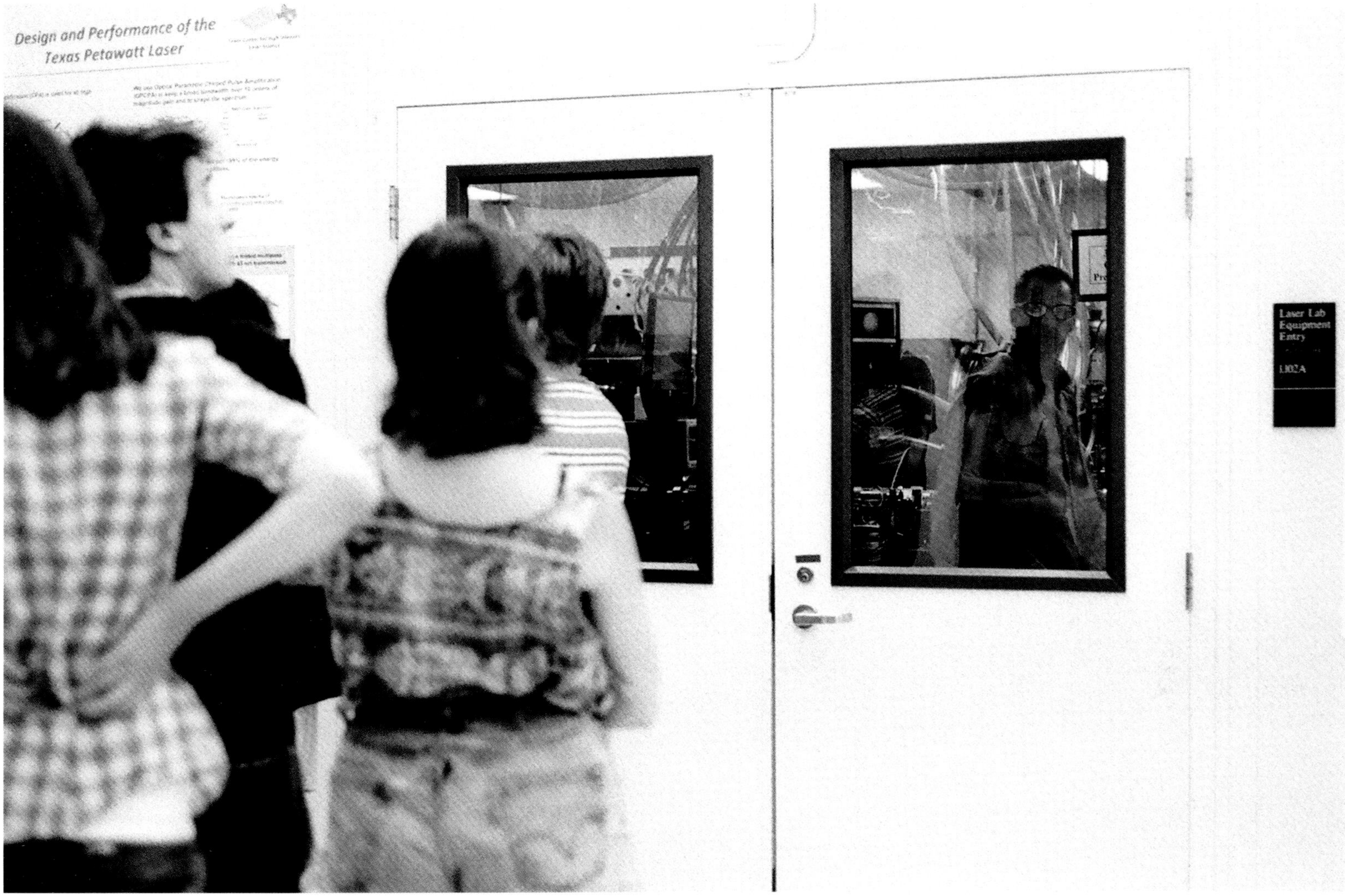
Design and Performance of the
Texas Petawatt Laser
Laser Lab
Equipment
Entry
L102A

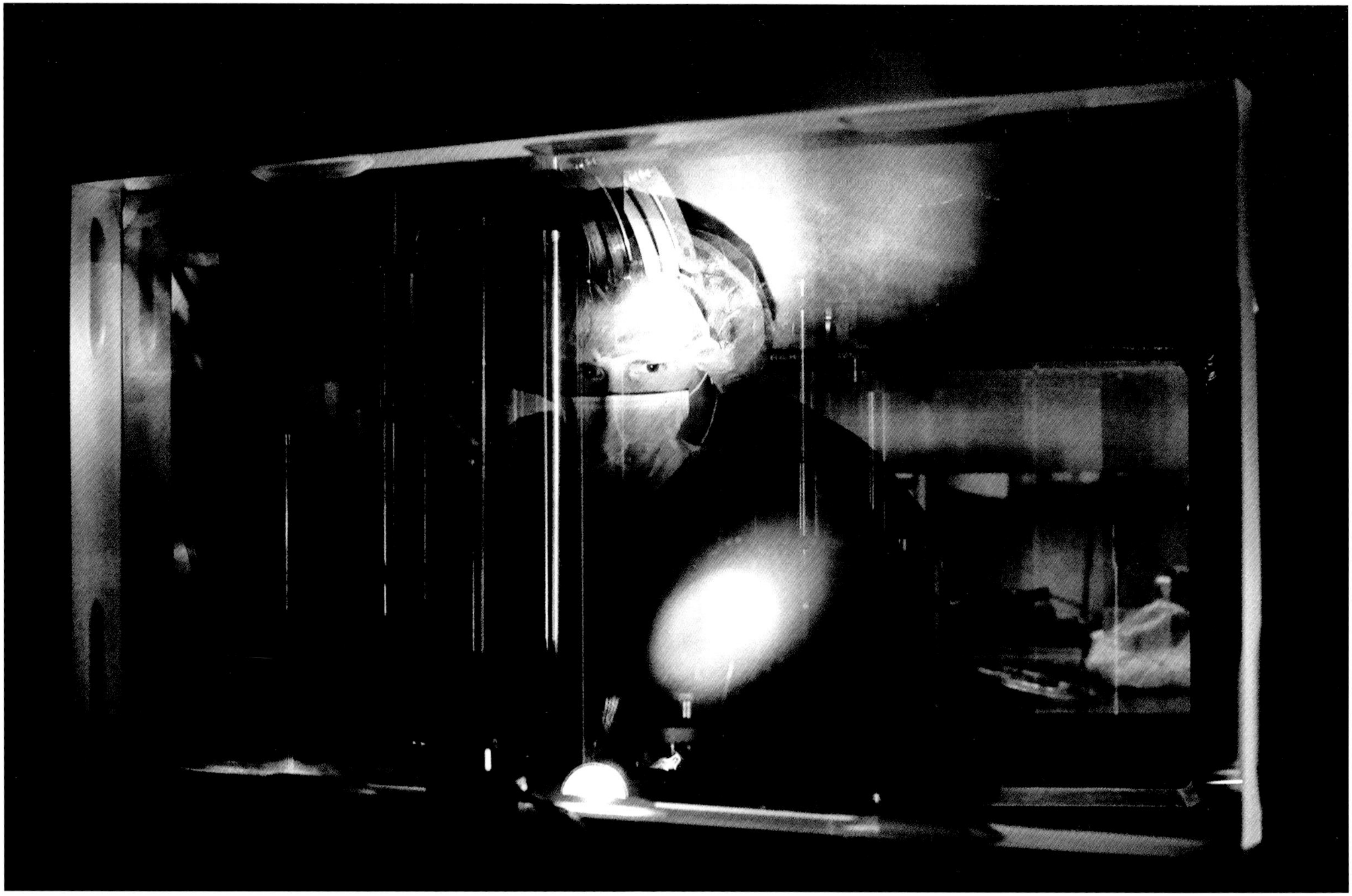

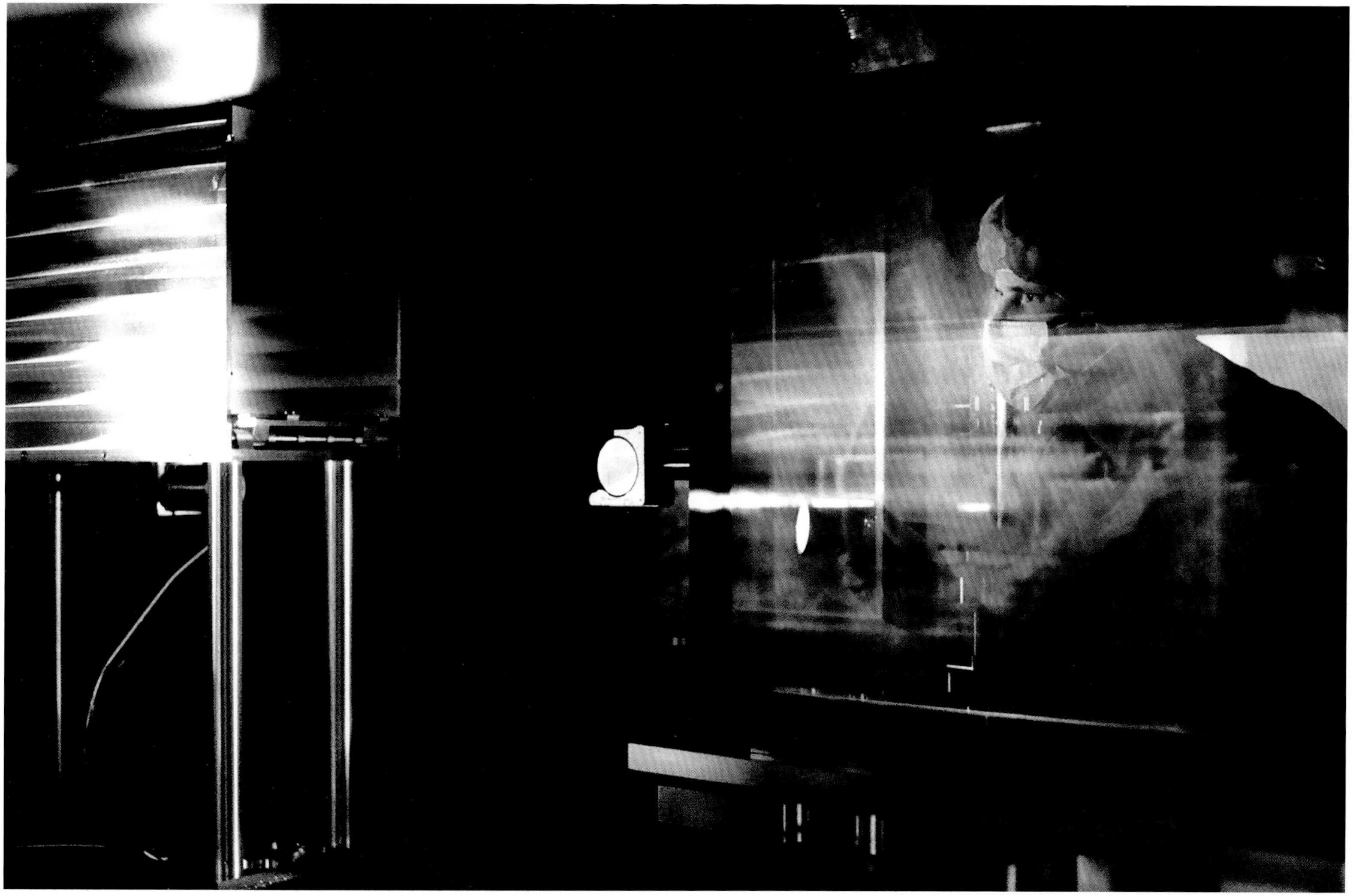

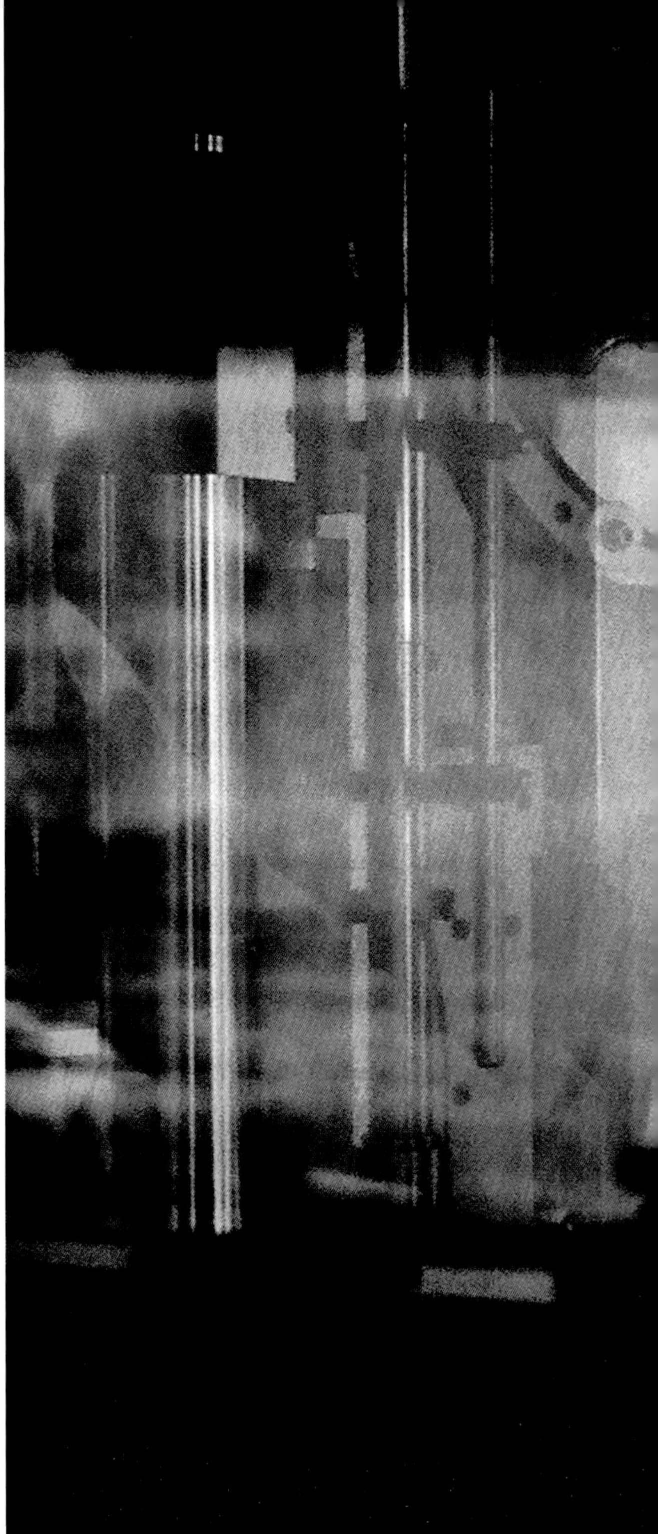

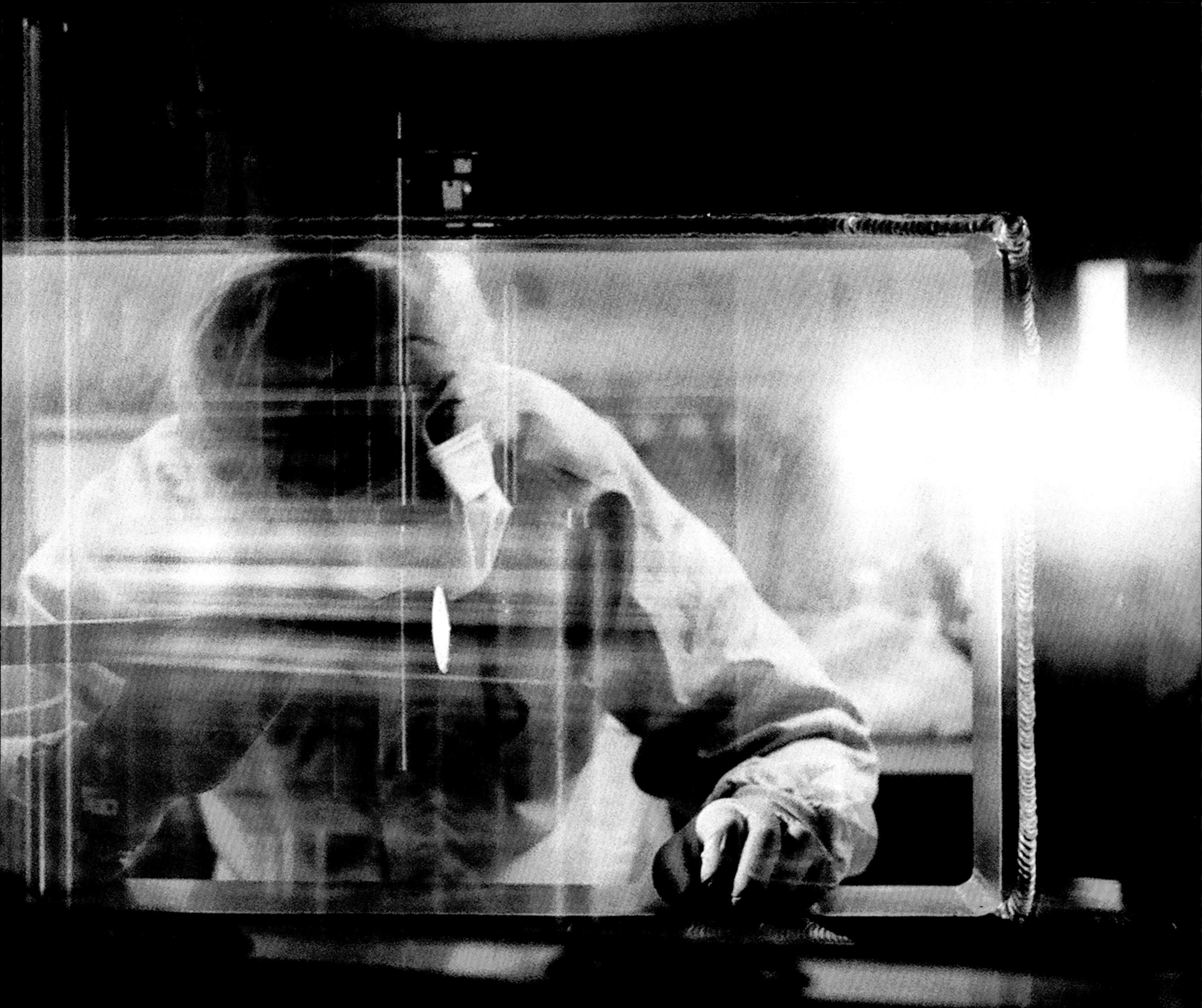

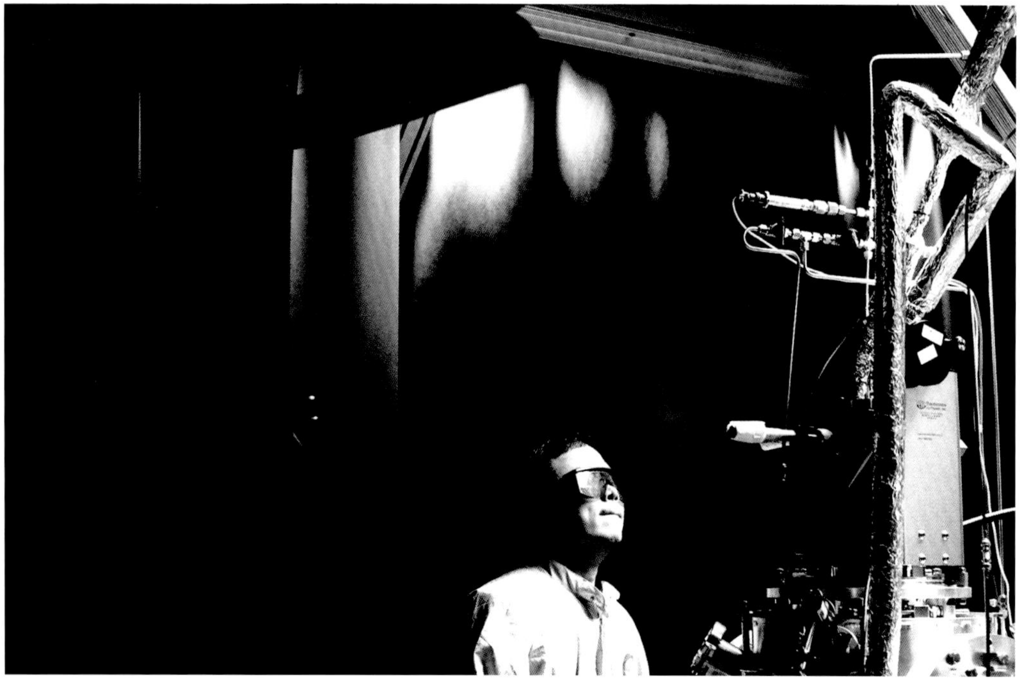

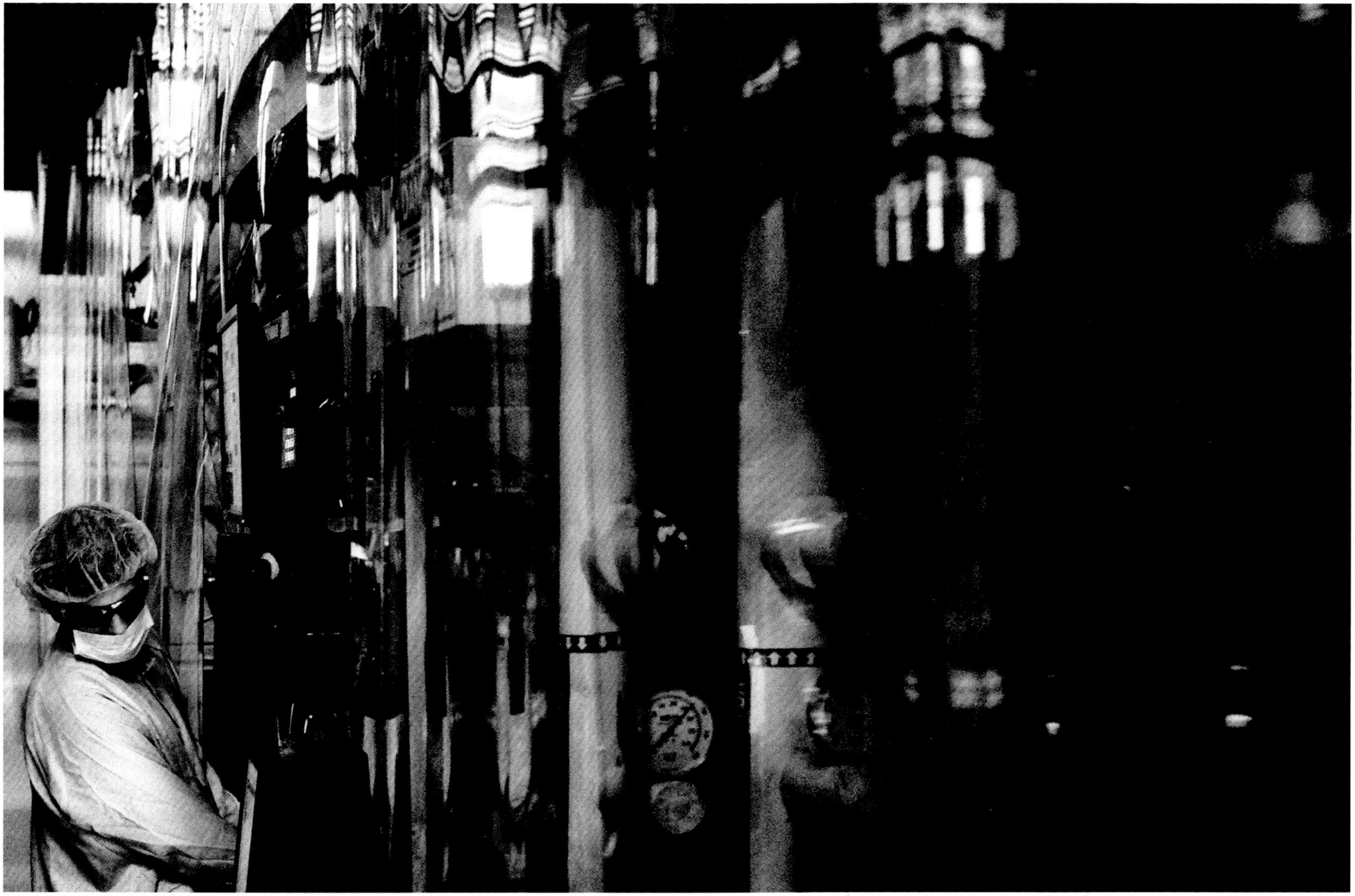

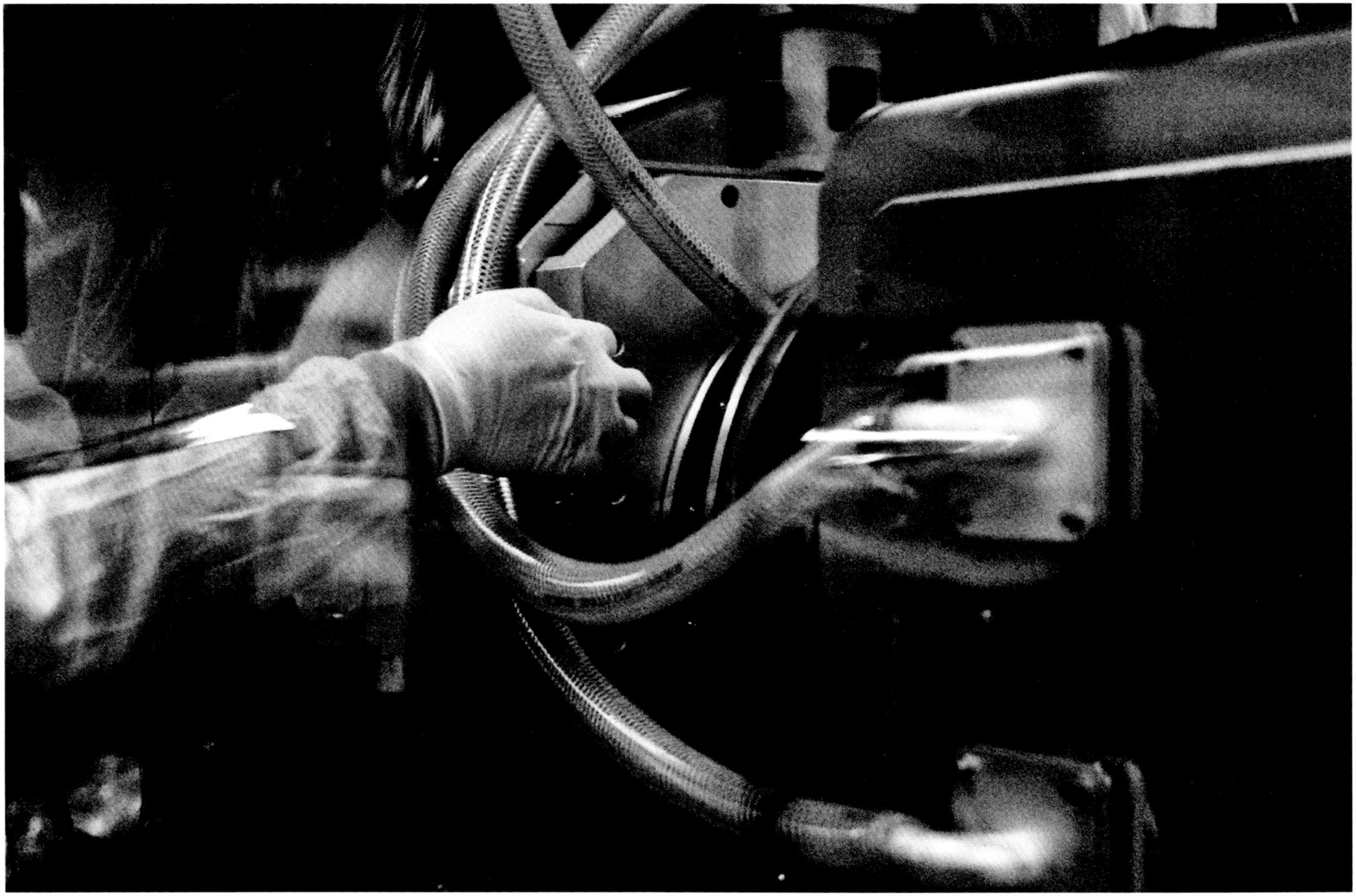

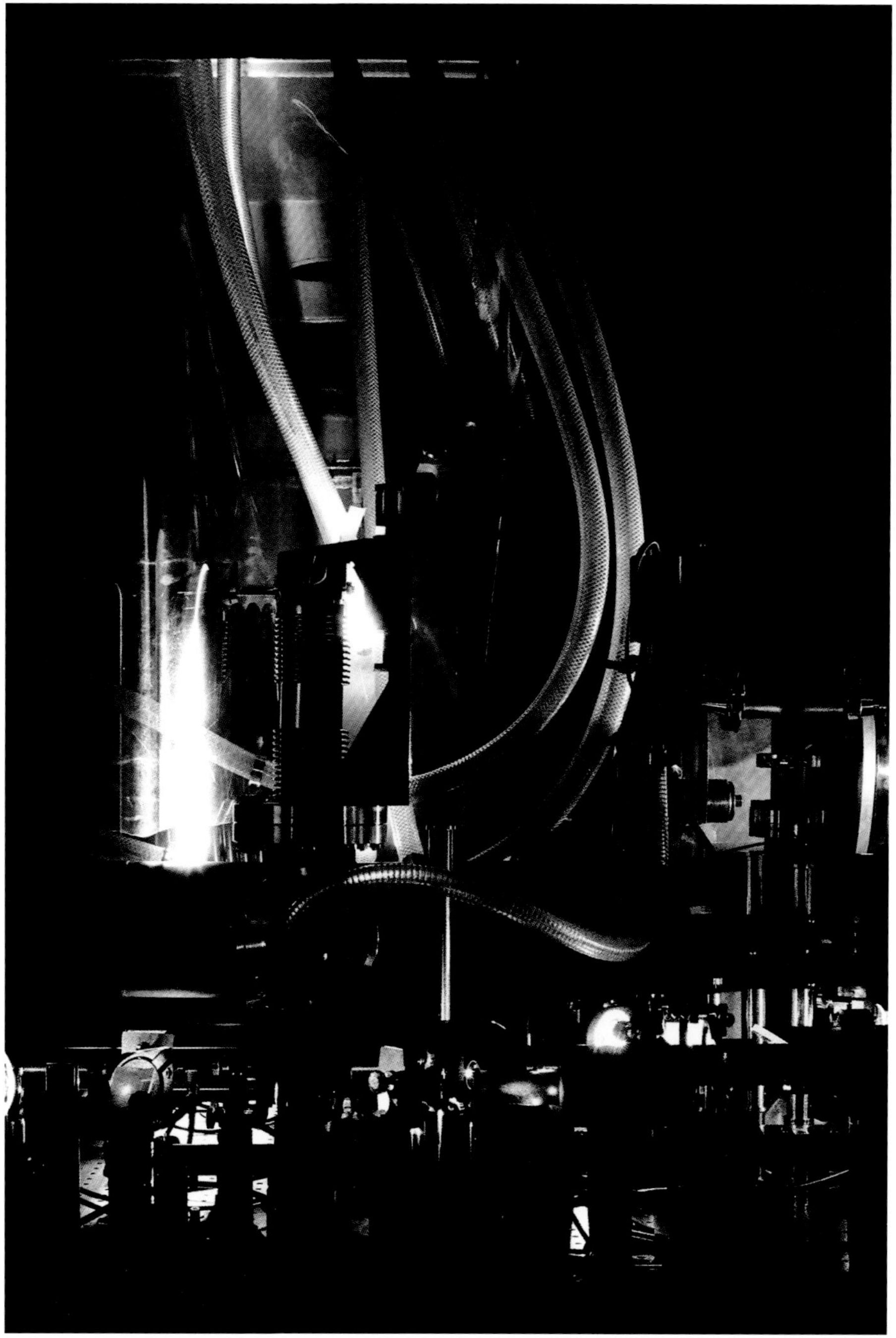

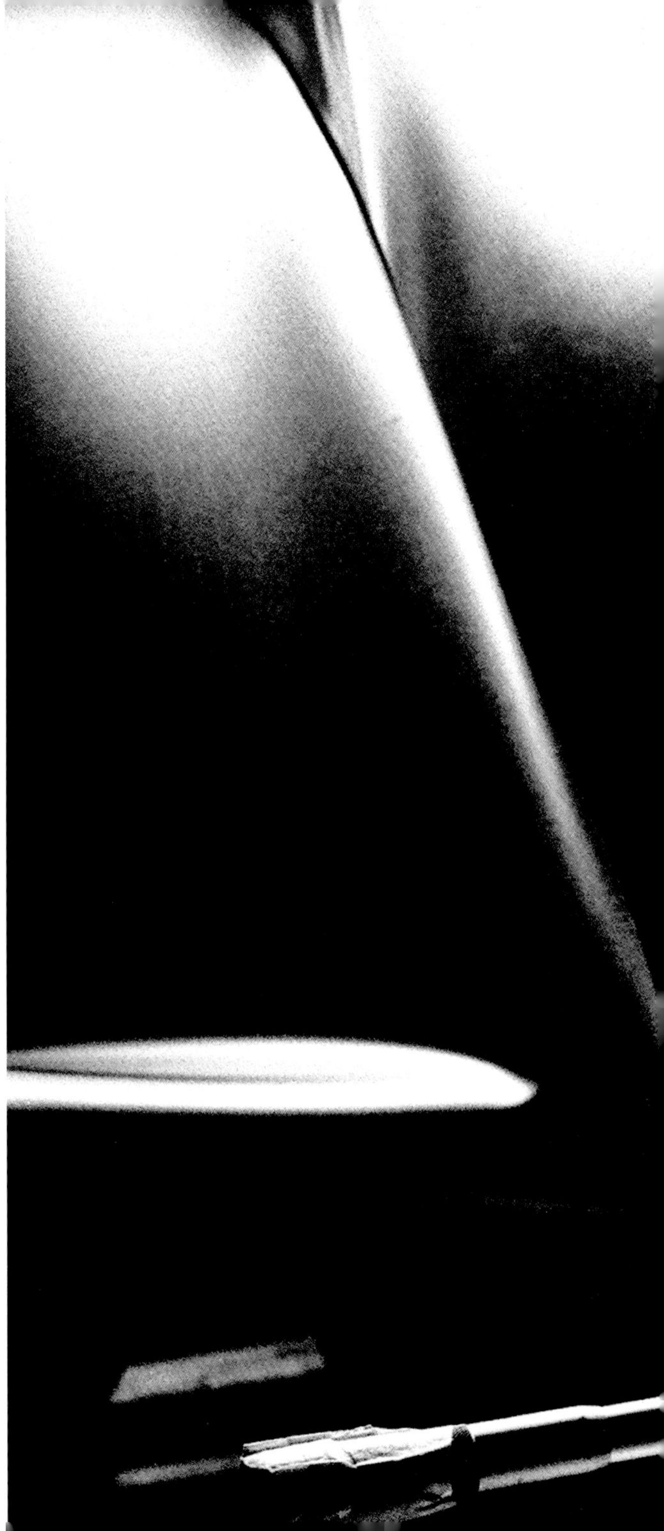

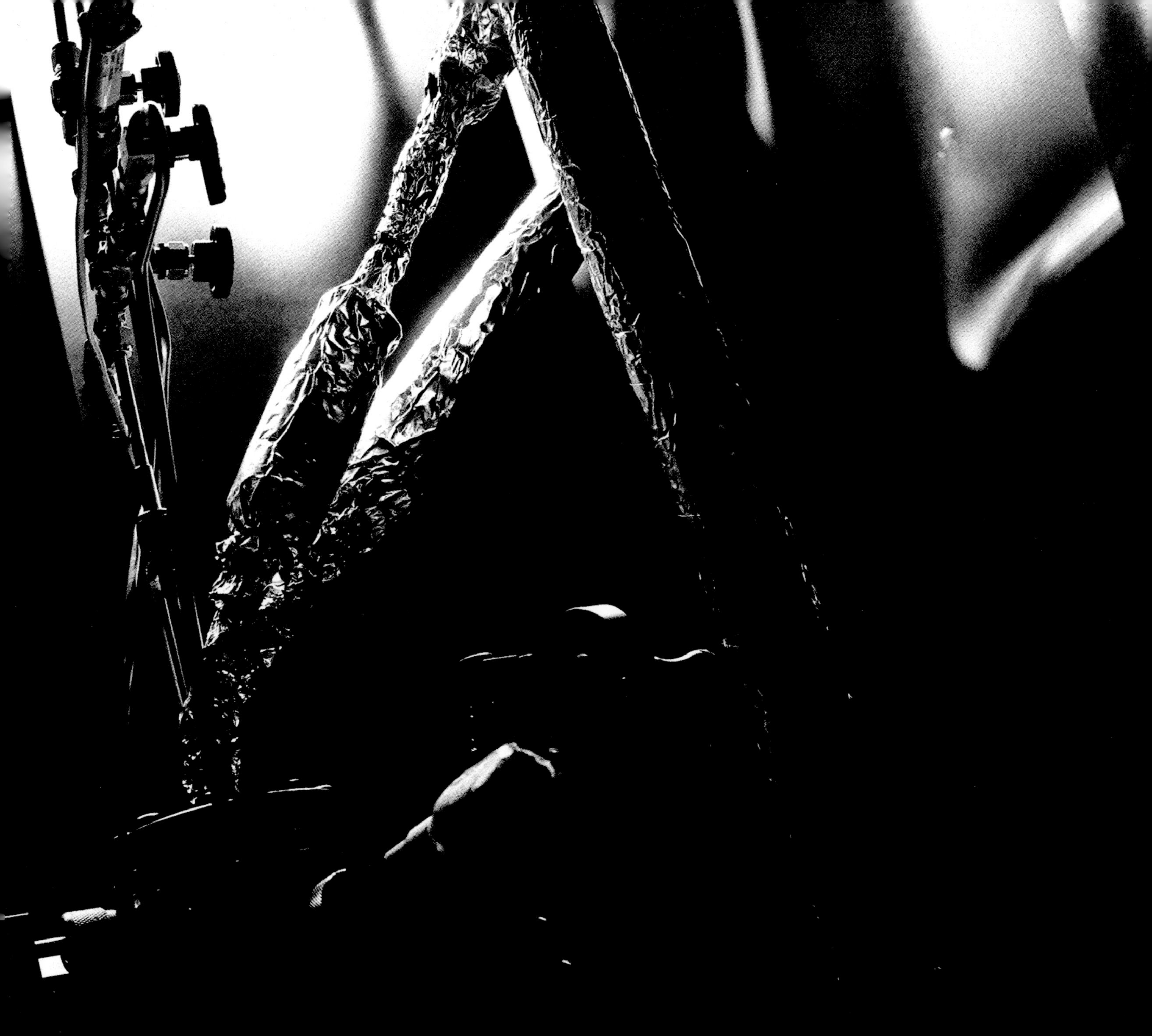

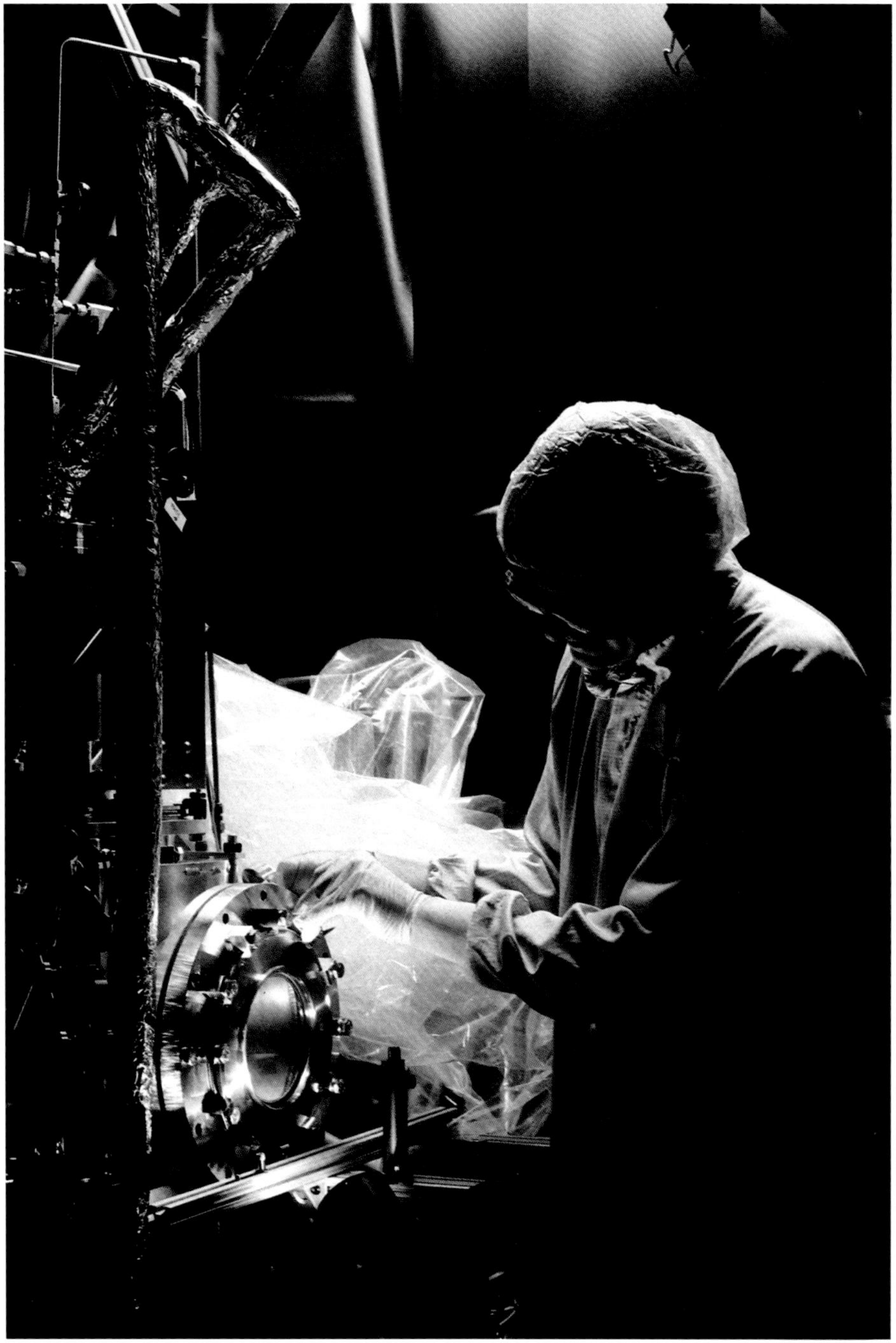

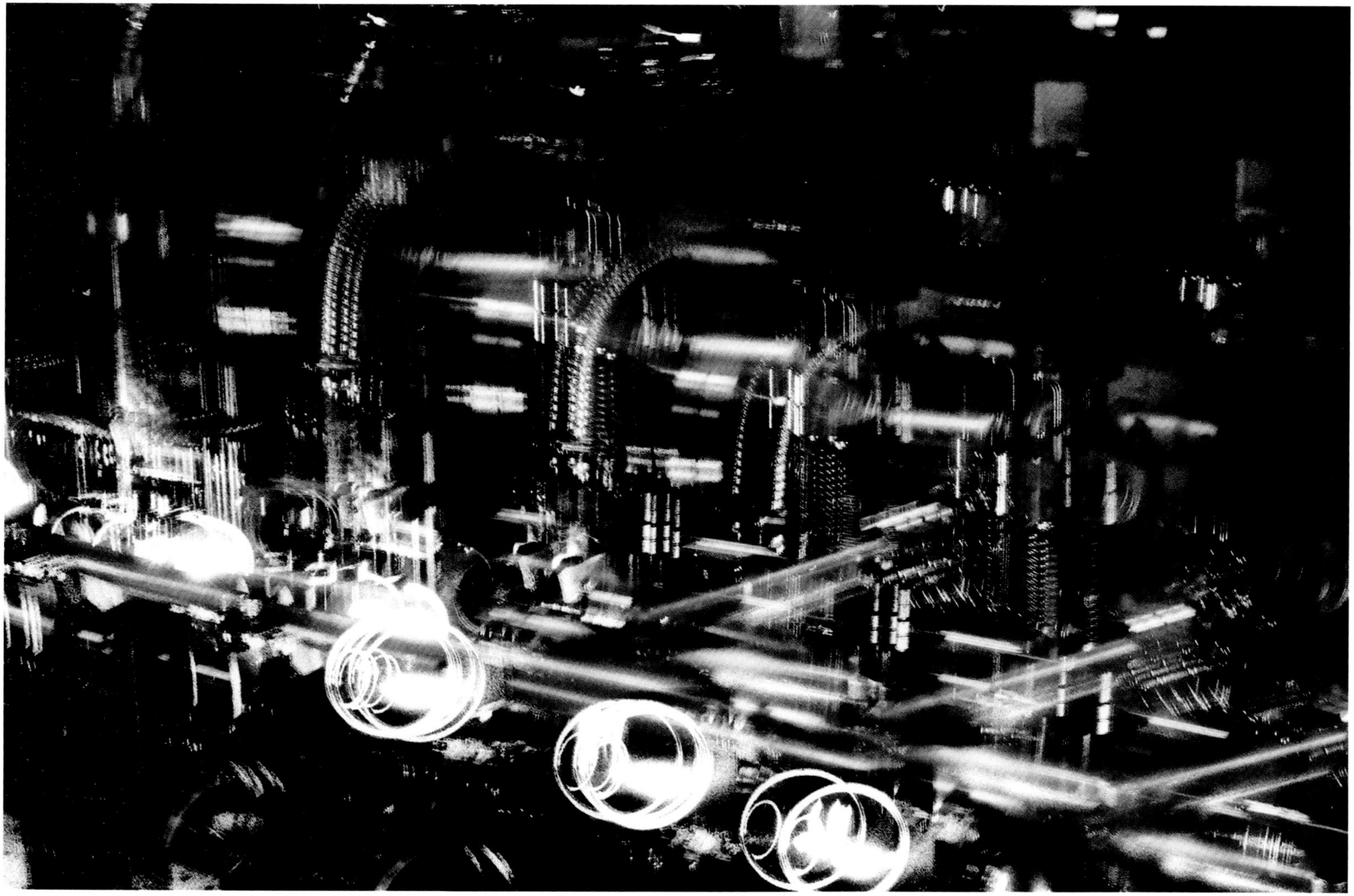

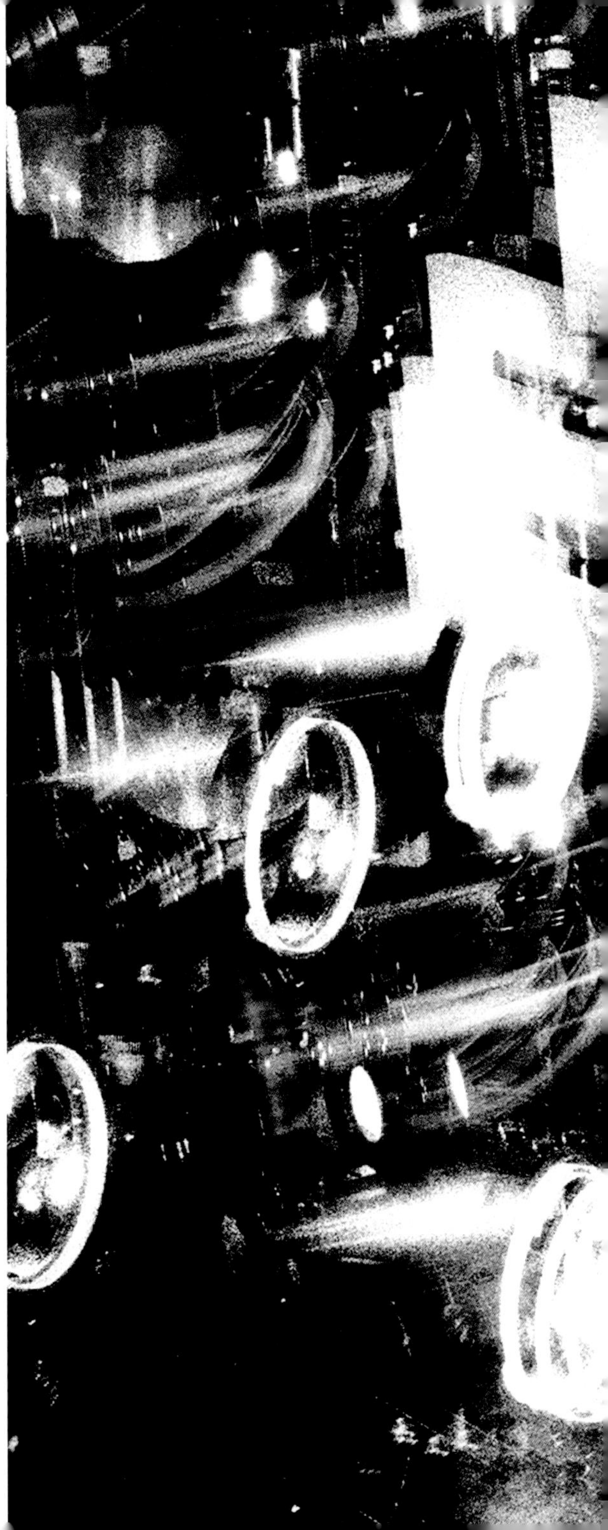

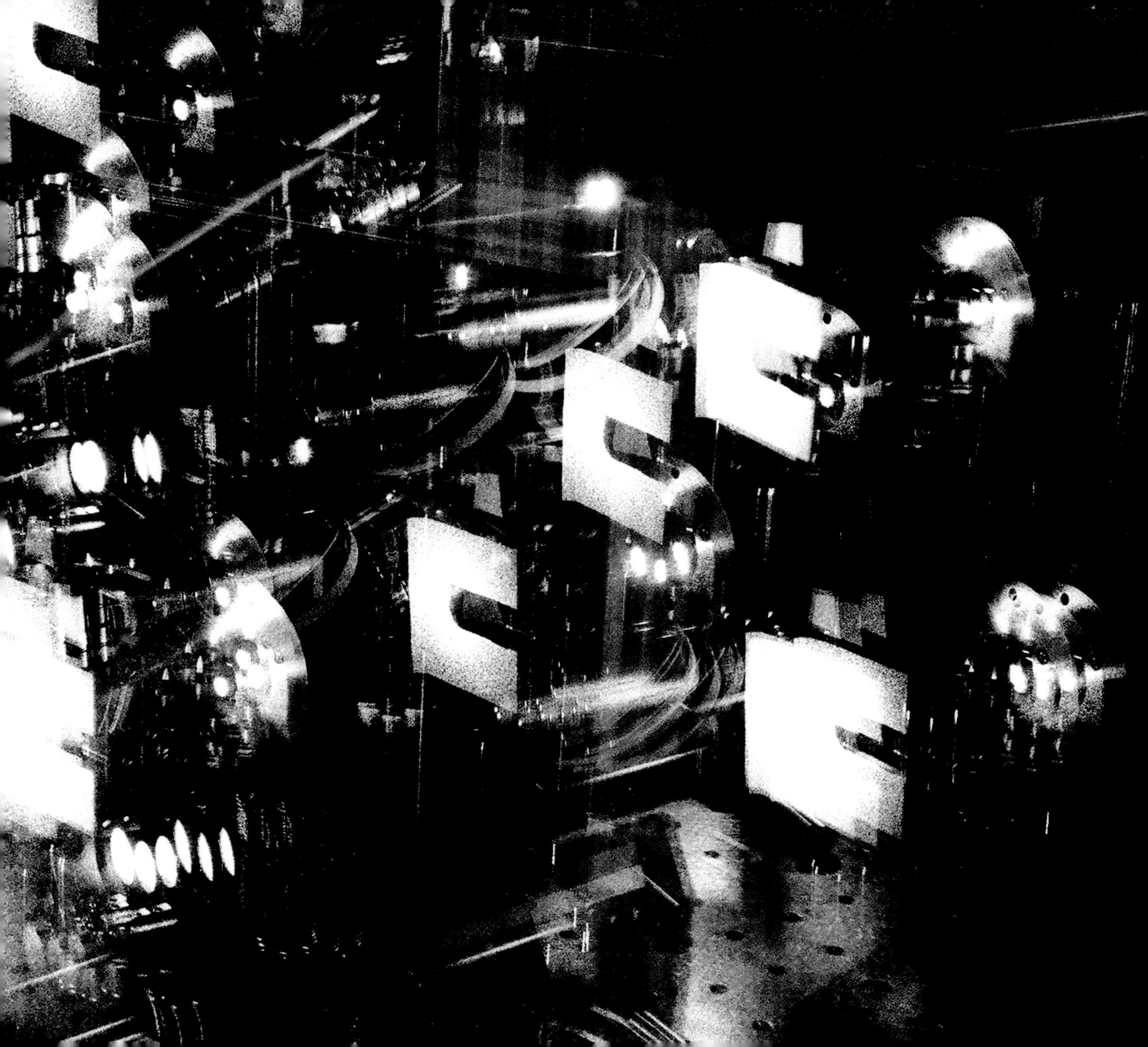

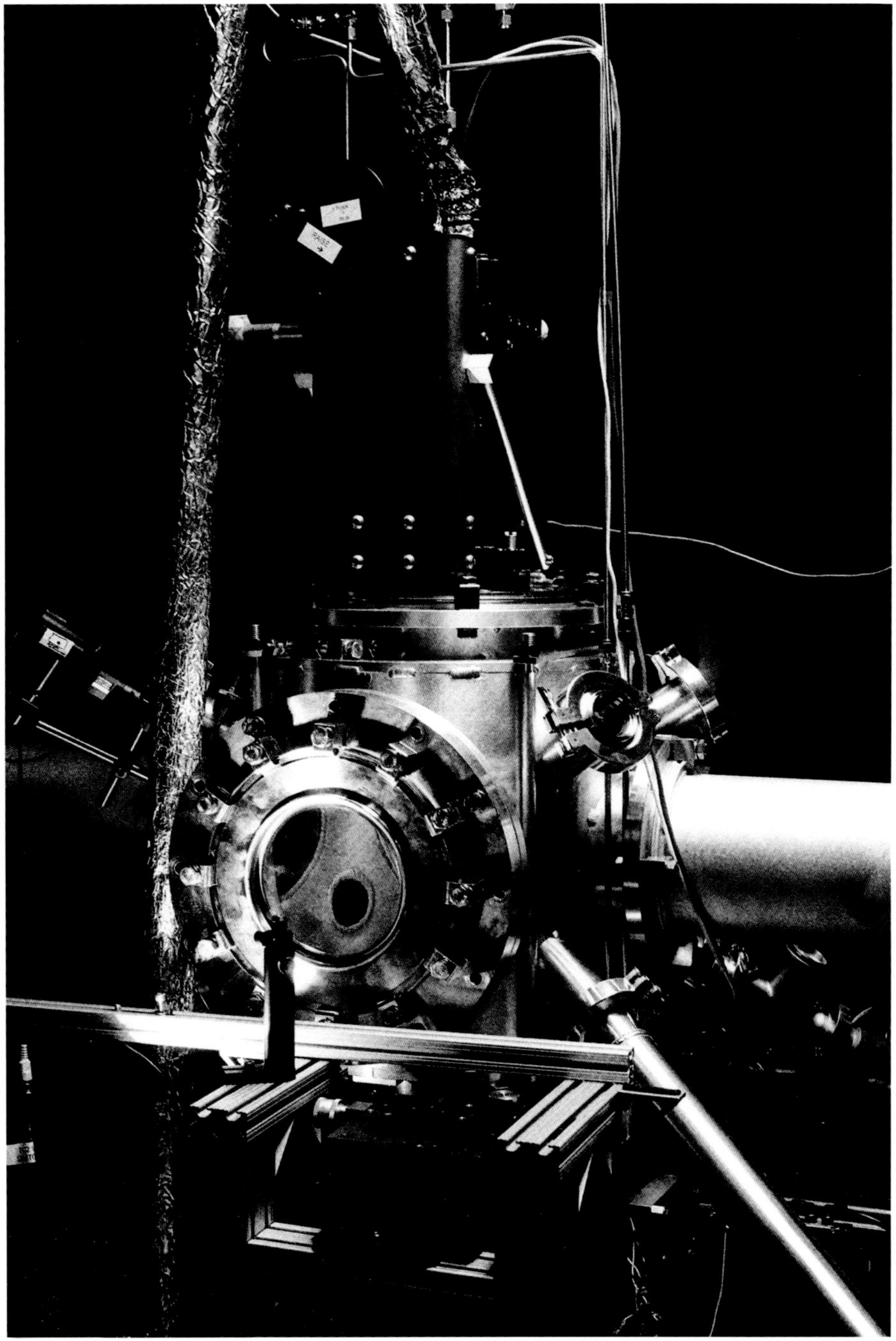

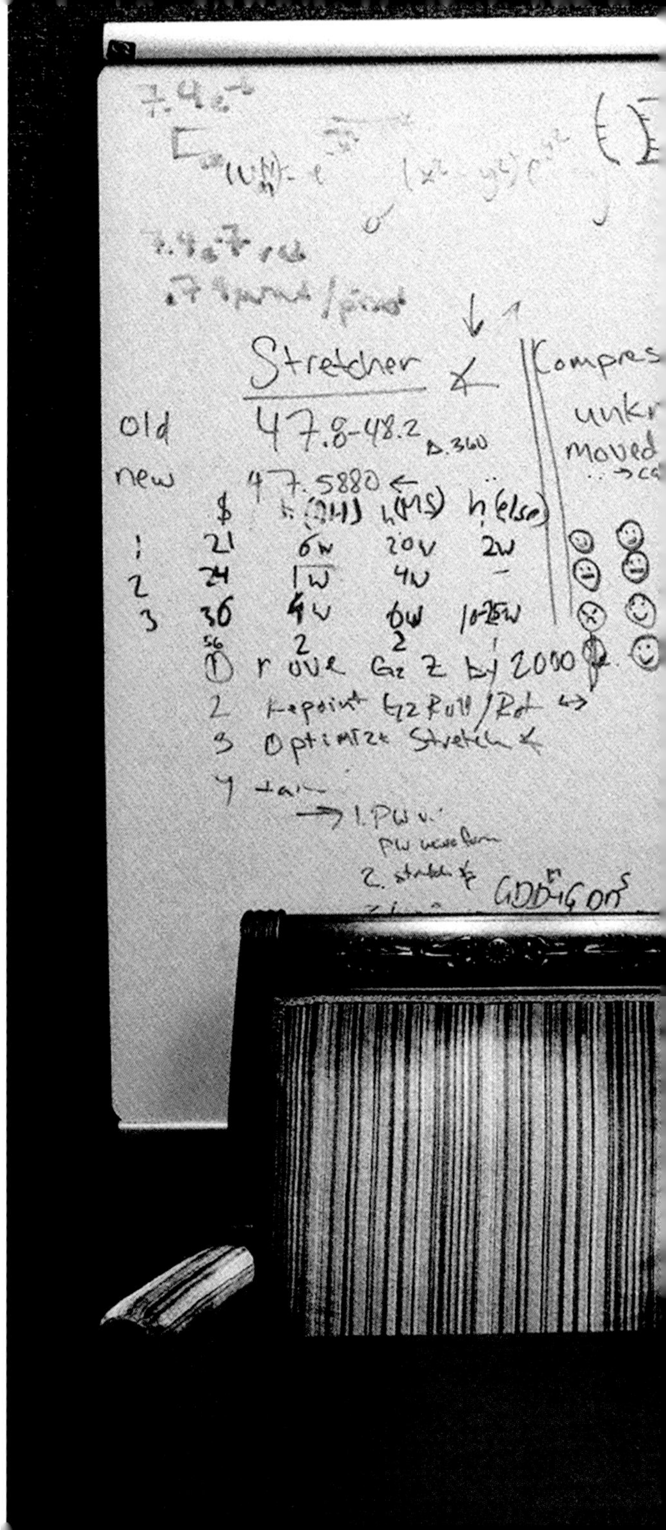
Stretcher
old
47.8-48.2
new
47.5880
Optimize Stretch

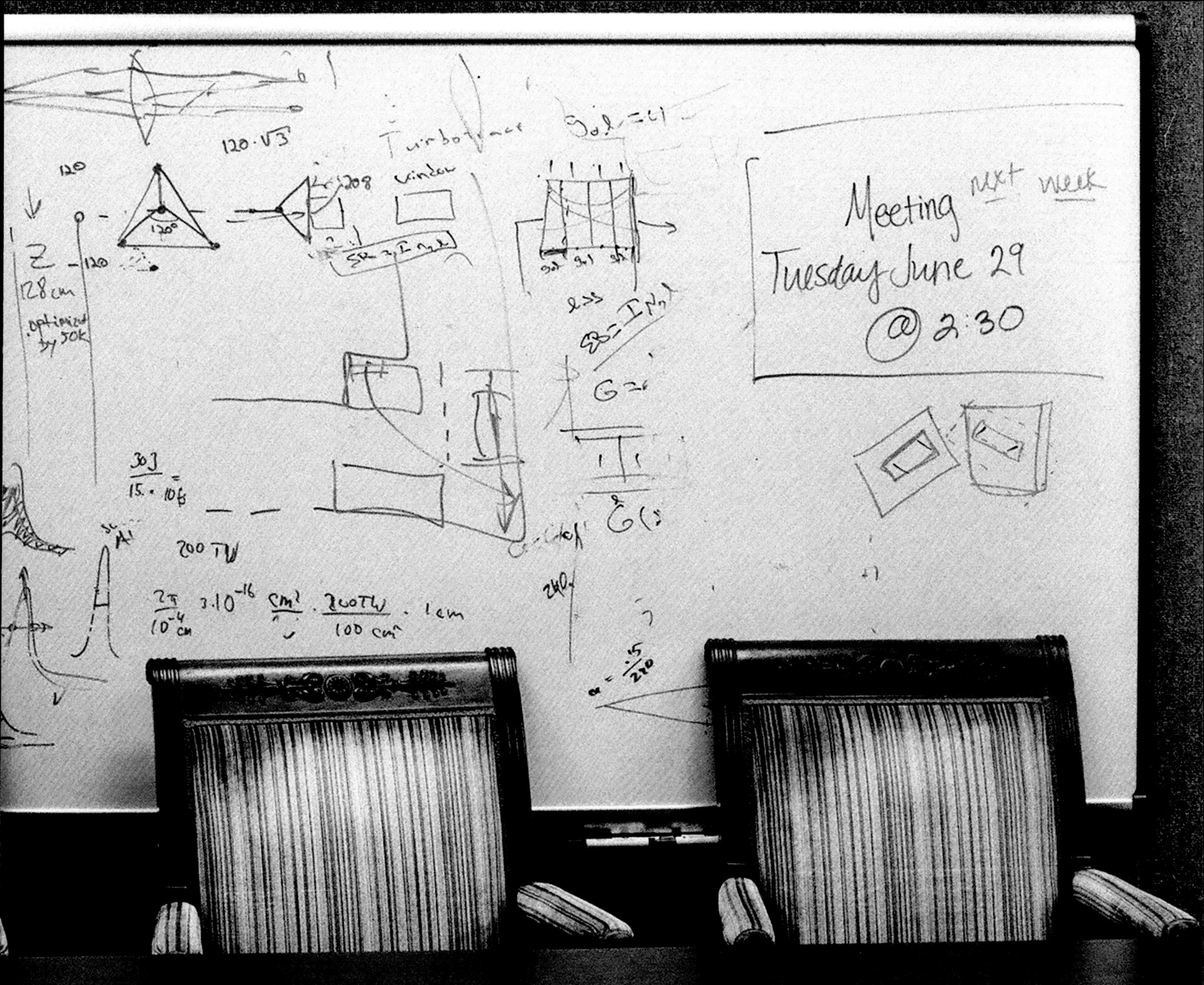
Meeting next week
Tuesday June 29
@ 2:30
120·√3
window
128 cm
200 TW

Acknowledgments

This book simply would not exist without the extraordinary contributions of Gilliss Dyer and Ted Borger, my principal chaperones in the Texas Petawatt Laser lab and the chief protagonists of these photographs. Thank you both for all your hard work and trust.

I am deeply indebted to the other scientists and engineers who agreed to be photographed for this series: Woosuk Bang, Erhard Gaul, Doug Hammond, Ramiro Escamilla, and Mikael Martinez.

I would also like to extend my sincere gratitude to Franki Aymond, Lindsay Fuller, Andrew Lafferty, Martin Ringuette, and Michael Spinks for their valuable technical support.

I am incredibly grateful to the Texas Petawatt Laser management team of Todd Ditmire, Aaron Bernstein, Michael Donovan, and Mikael Martinez for their tireless advocacy and support.

My deepest thanks must be reserved for my wife and children—Tara, Scarlett, and Phoebe—for their apparently endless reserve of patience and encouragement during this protracted endeavor.

ACKNOWLEDGMENTS

This book has benefited immensely from the contributions of many generous benefactors, especially National Energetics, the laser design firm that has brought the Texas Petawatt Laser's core technology to the world at large.

I am inexpressibly grateful for the support of Michael Barnstijn, Dr. George A. Condas, and the George and Cheryl Lauterstein family, and I am inspired by their respective achievements in engineering, physics, and photography.

I am particularly indebted to Robert B. McLeod, as well as Robert and Amelia Brenner and Lauren DeLong for their humbling efforts to bring this book to life.

This project would not have seen publication without the kind support of Charlene Gilliss Dyer, Frank J. Tilicek, David Stewart Zink, Rev. Ronald G. Cosseboom, Jeff and Barbara Goodman, Jennifer M. Watters, Durell D. Agha, and Darrell Gum.

I would also like to express my sincere gratitude to Baptiste Bentivoglio, Rex Aymond, Mojo Norman, Jon Paul Buchmeyer, Juan Pablo Chavez, Dave Dall, Nancy Haug, David Lykes Keenan, Nikhil Ramkarran, Pamela Buchmeyer, Shellie Crandall, Nancy Kickertz, Gay Dearing, Ryan Winkler, Timothy Herr, and Donald and Ginette Brenner.

I am deeply thankful for the boundless creativity and support of Michael Itkoff, Taj Forer, Ursula Damm, Laura Fredrickson, and the entire team at Daylight Books.

Lastly, I am honored and humbled by the incredible contribution that Dr. Rudy Rucker has made to this volume.

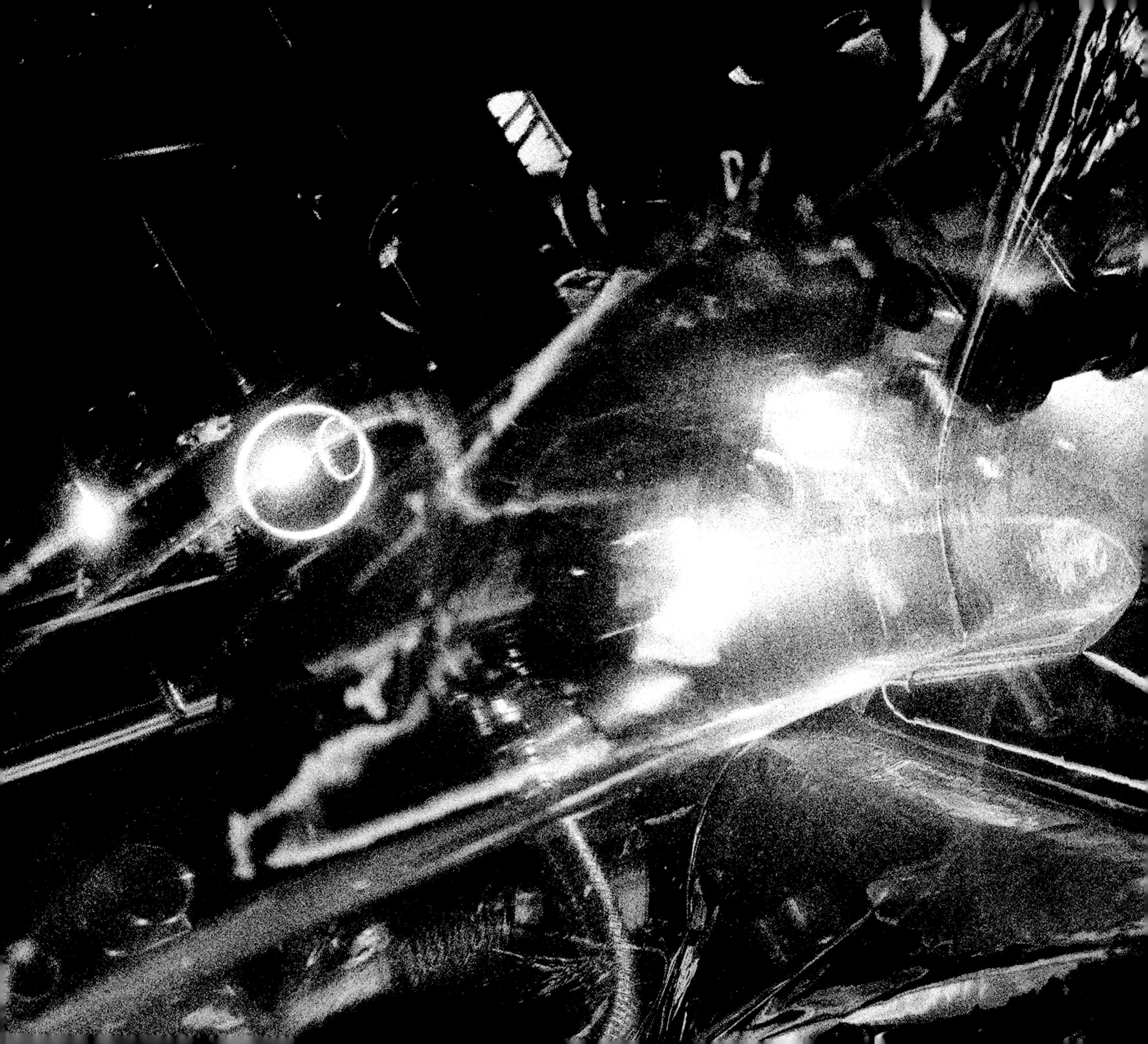

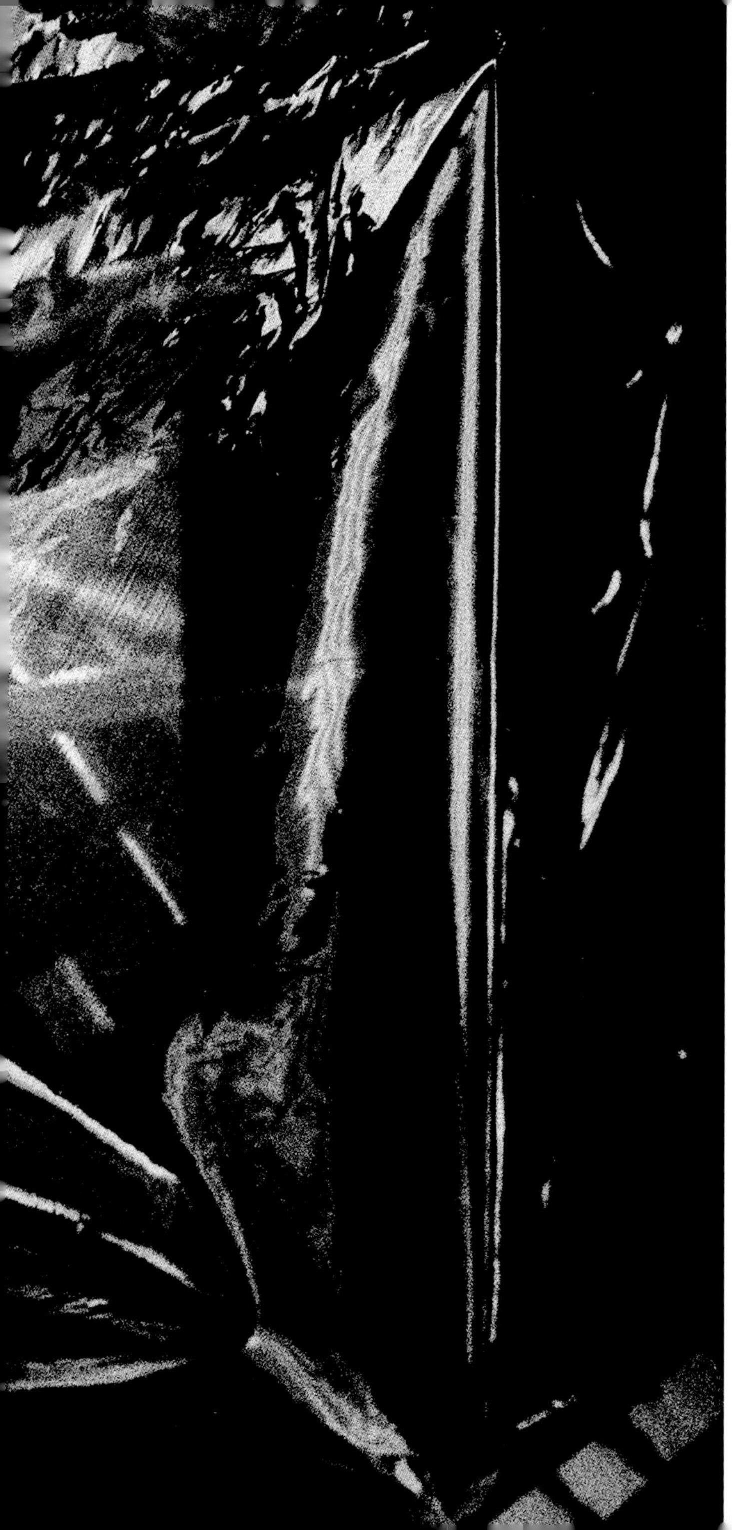

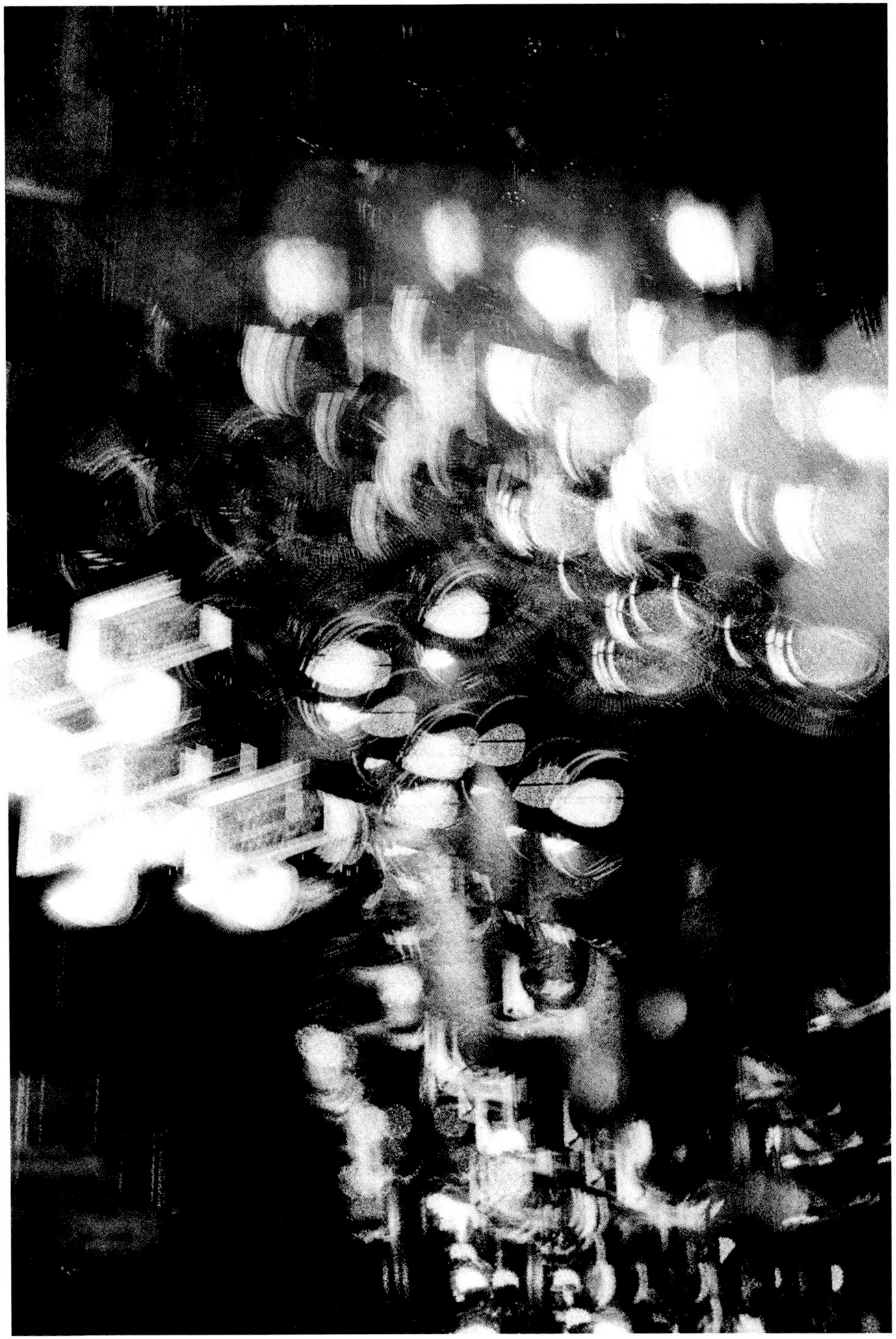

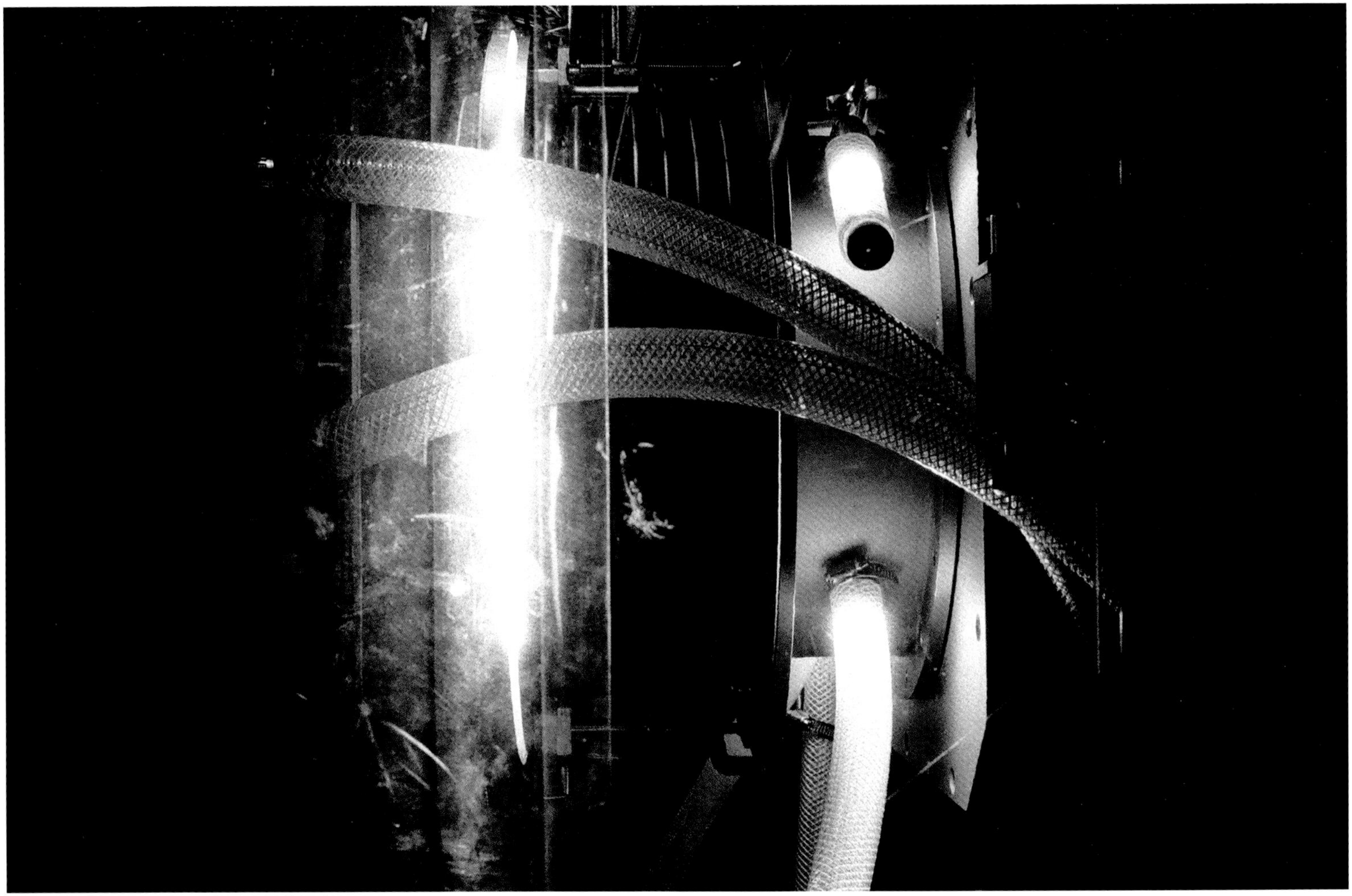

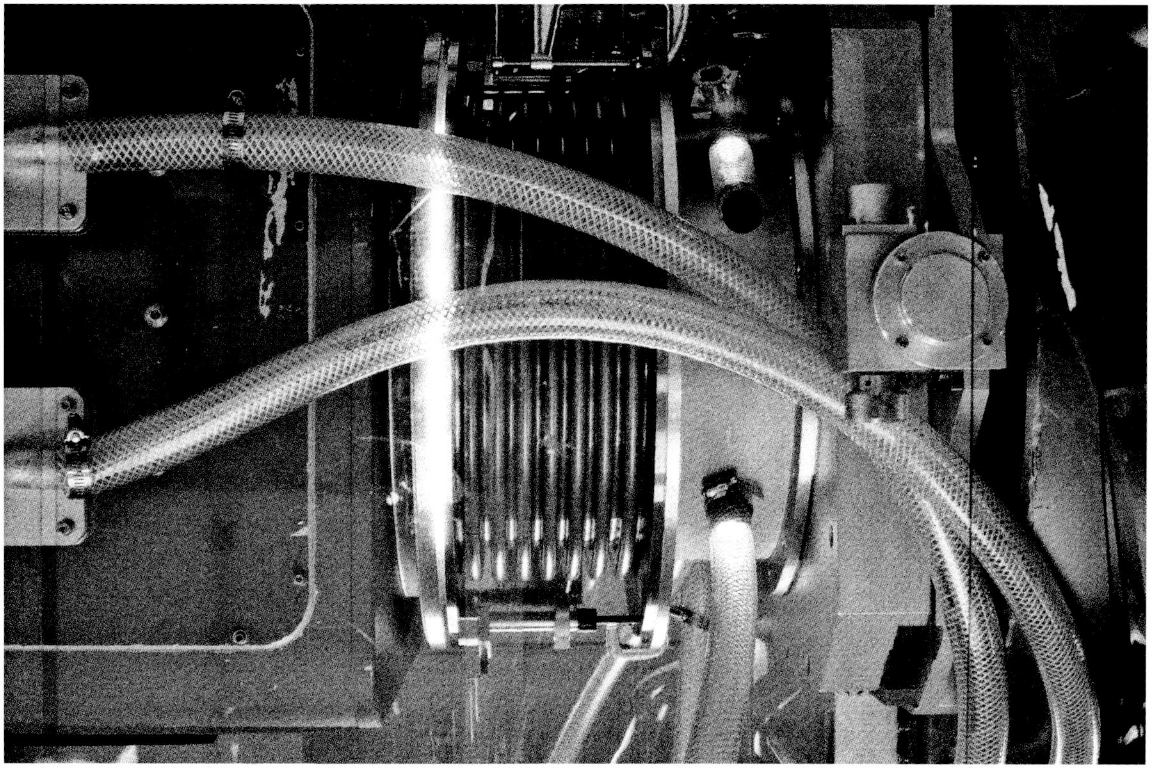

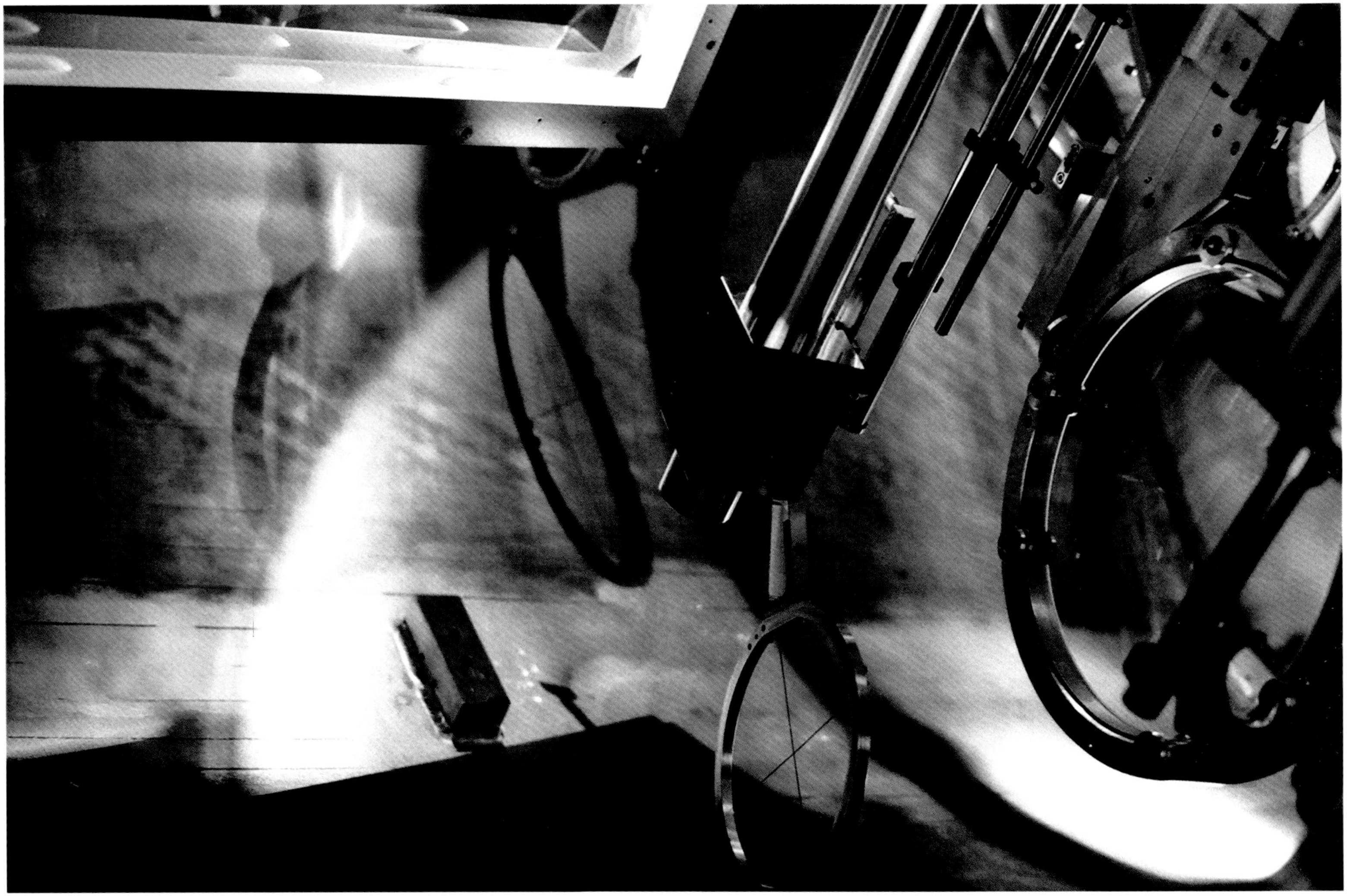

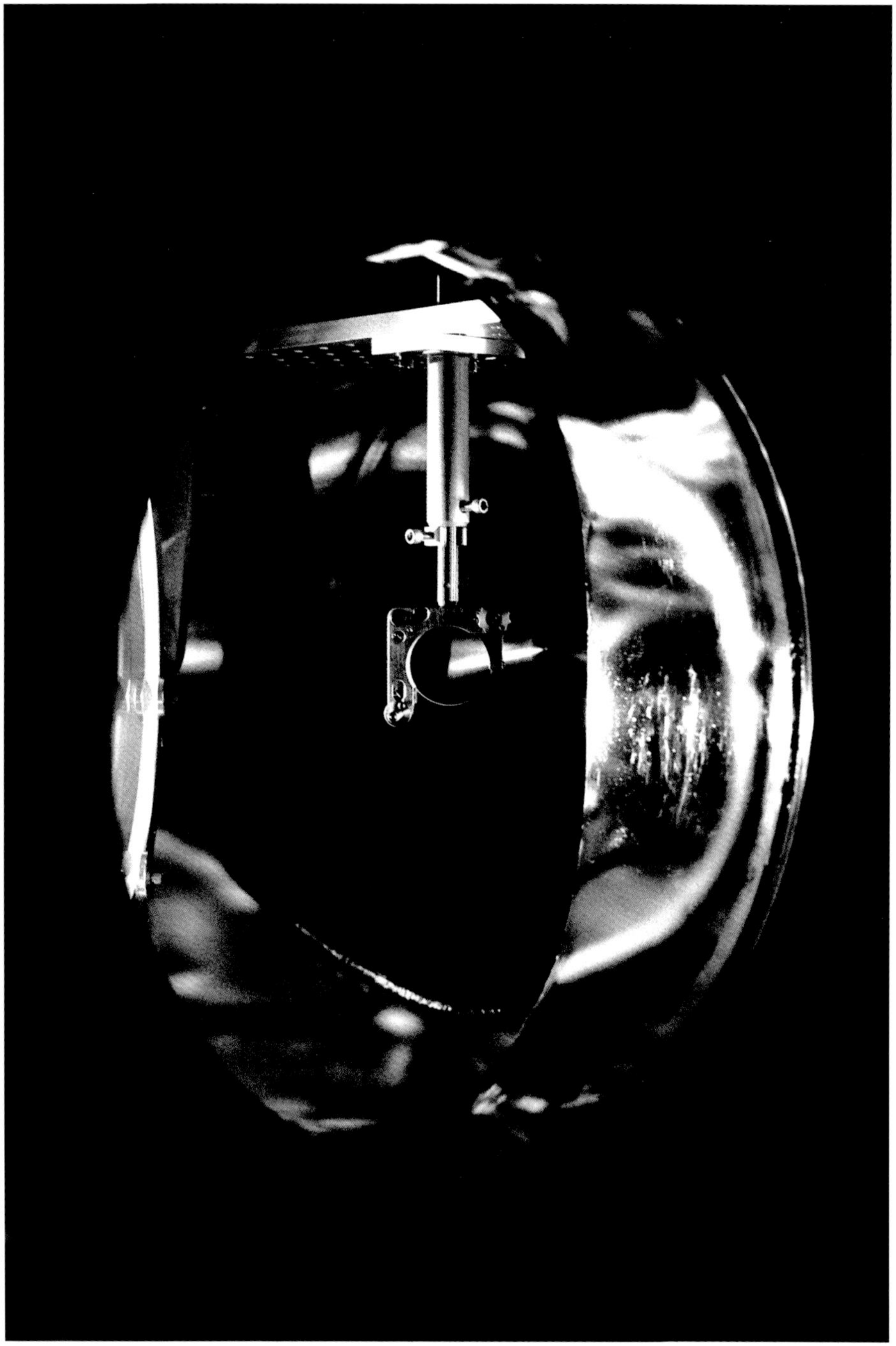

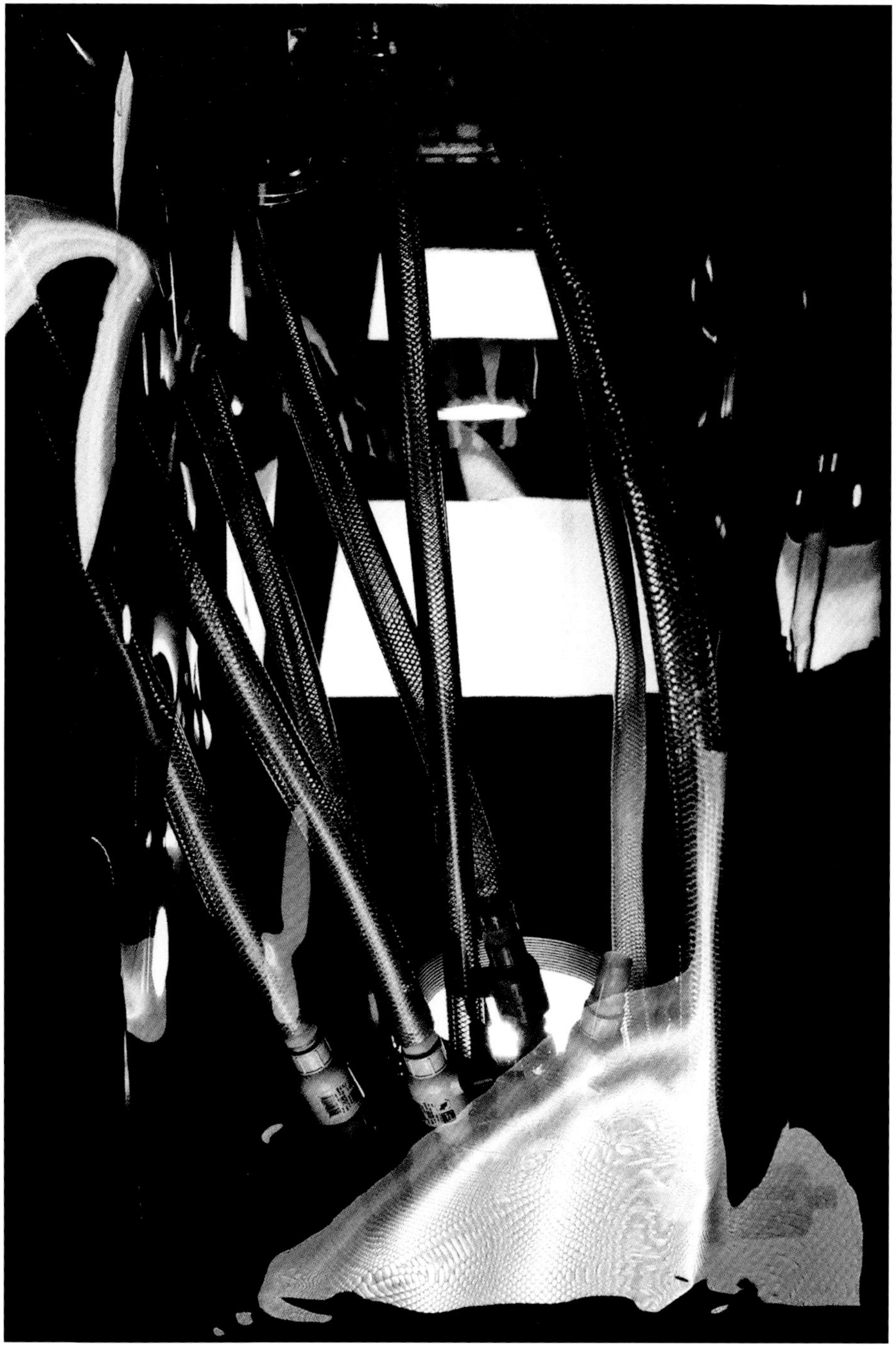

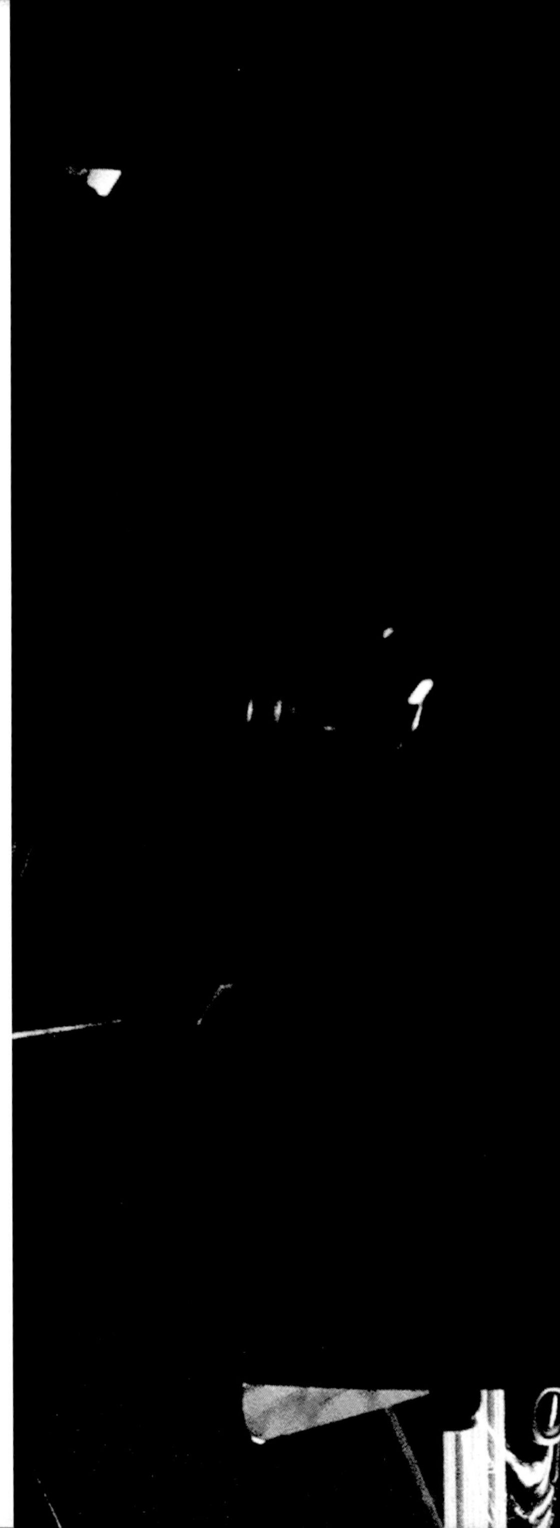

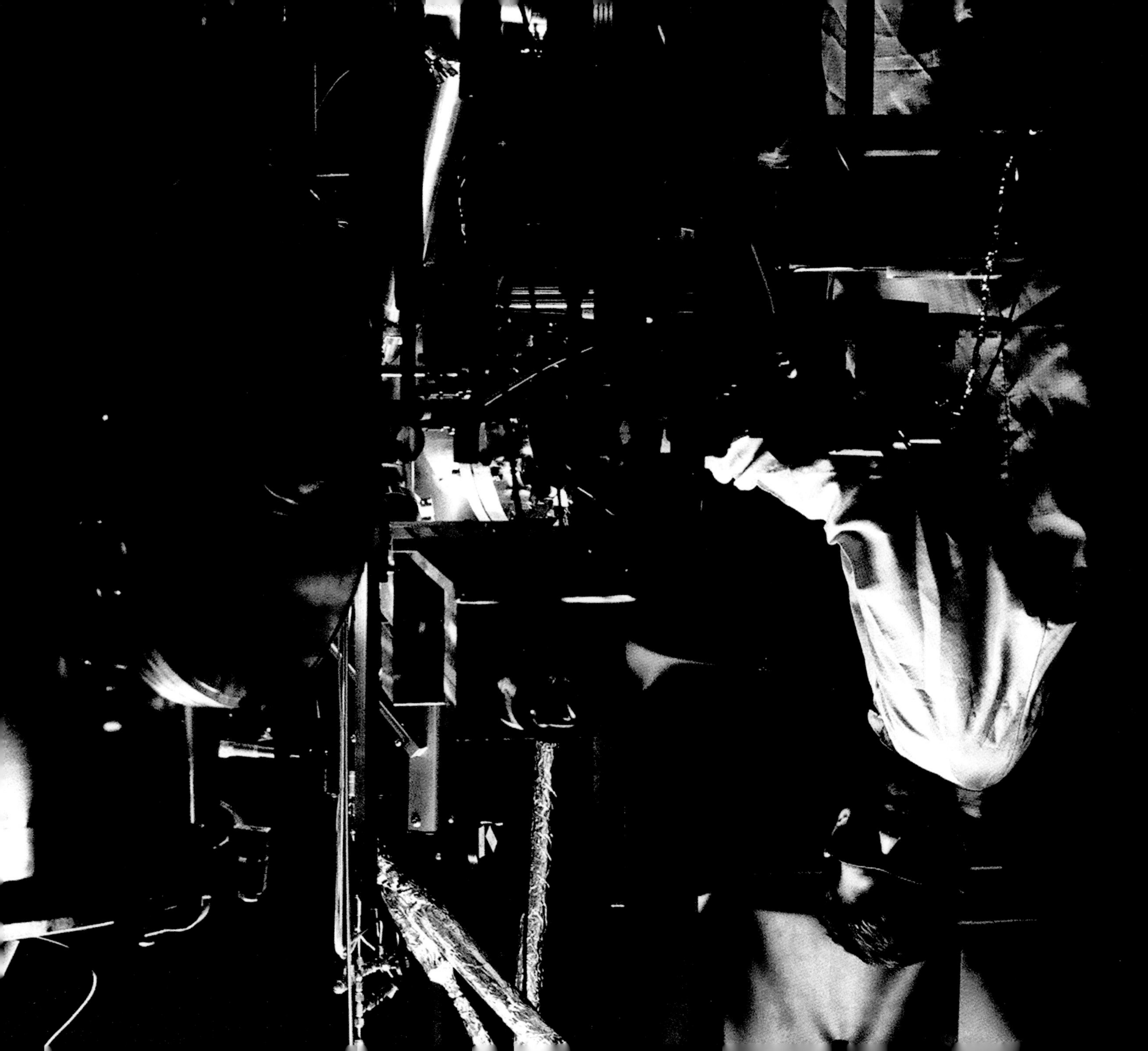

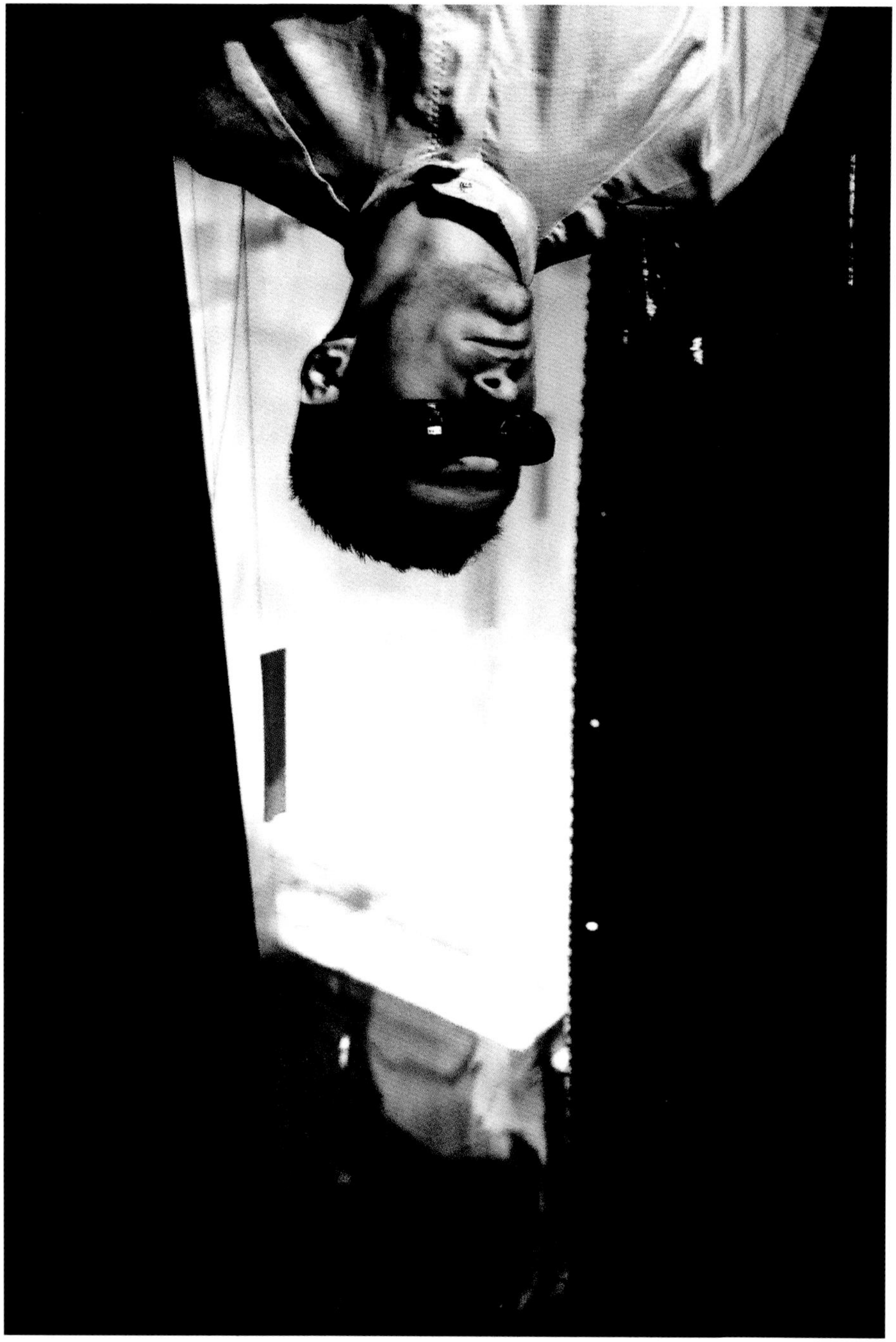

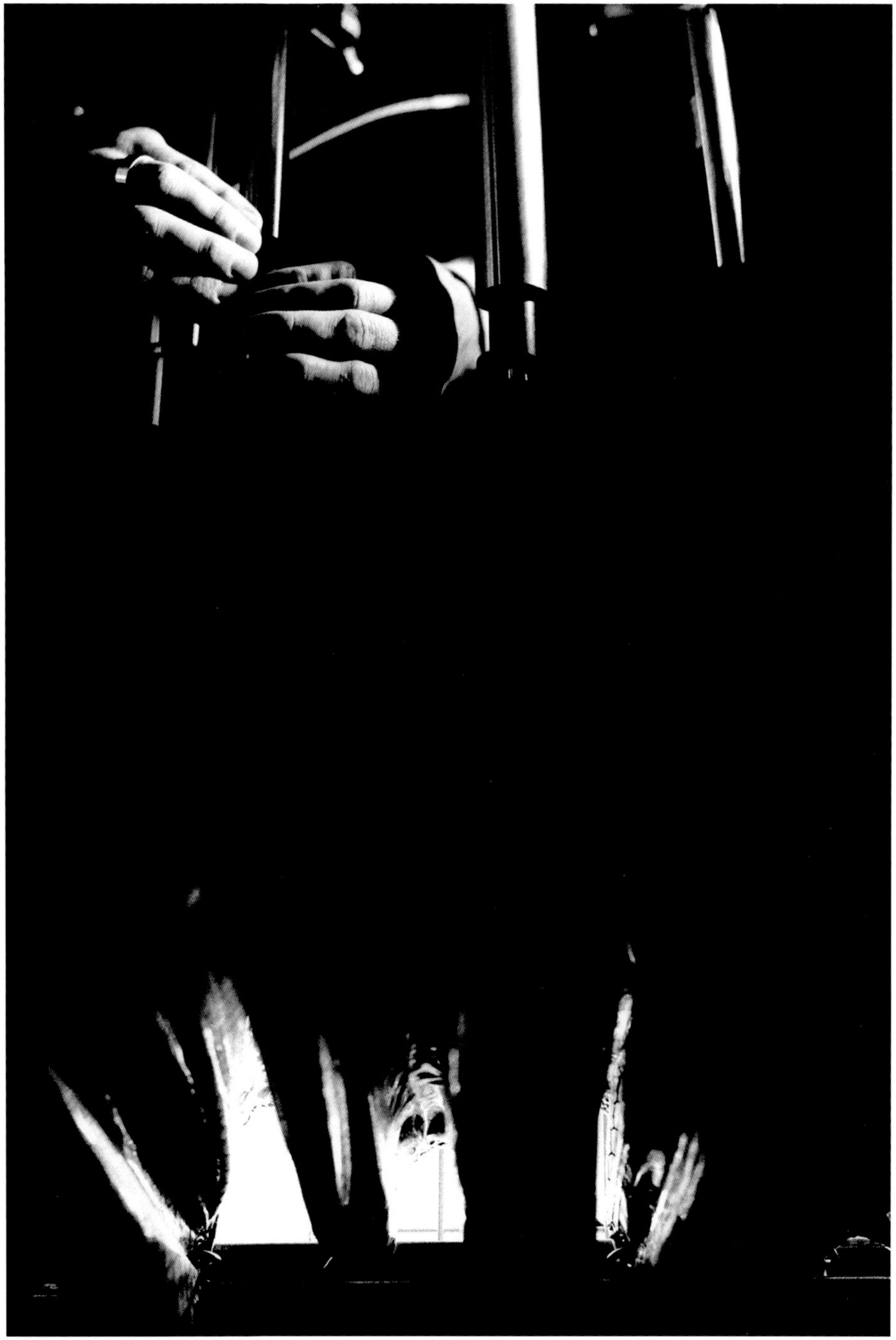

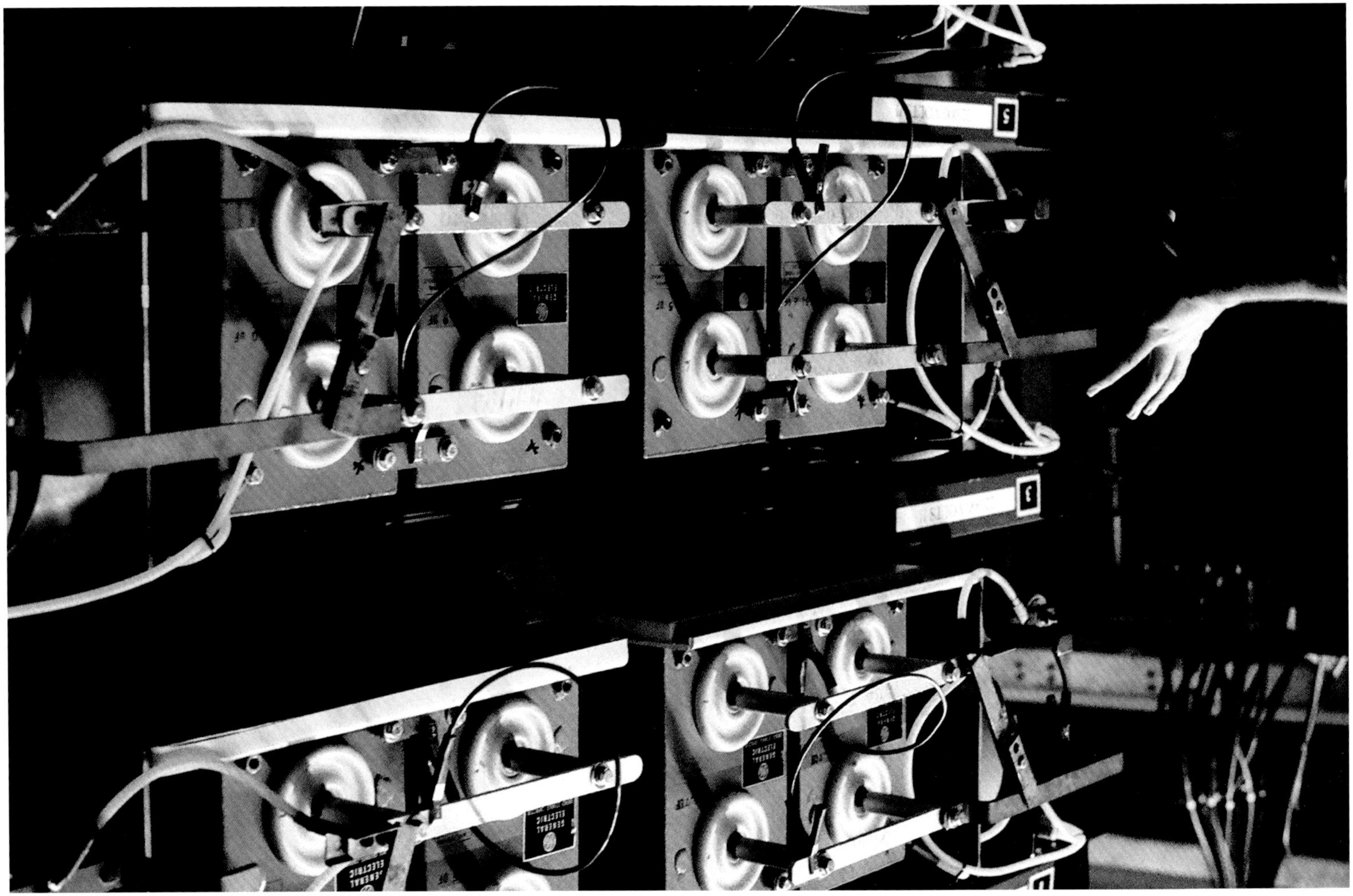

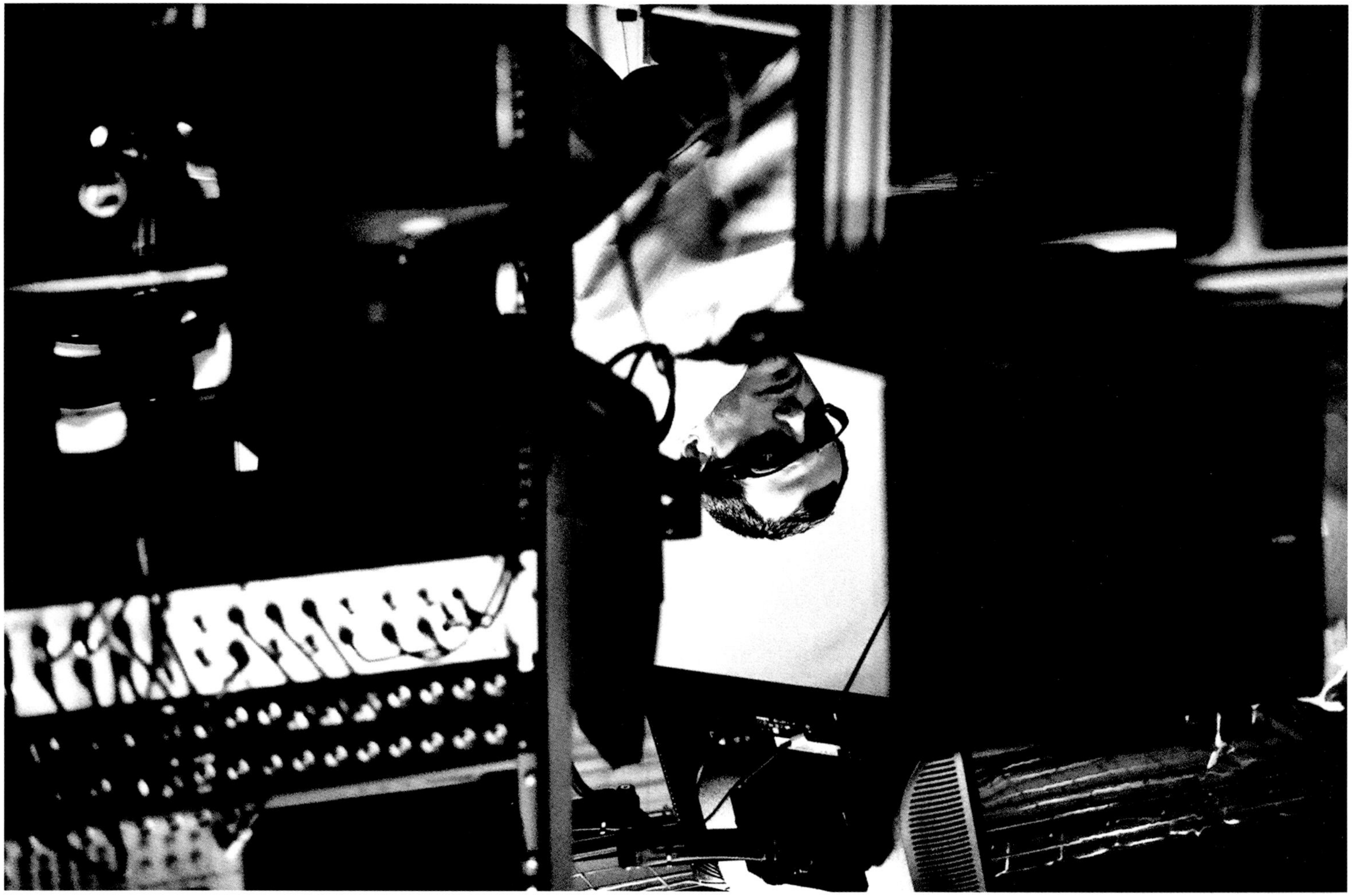

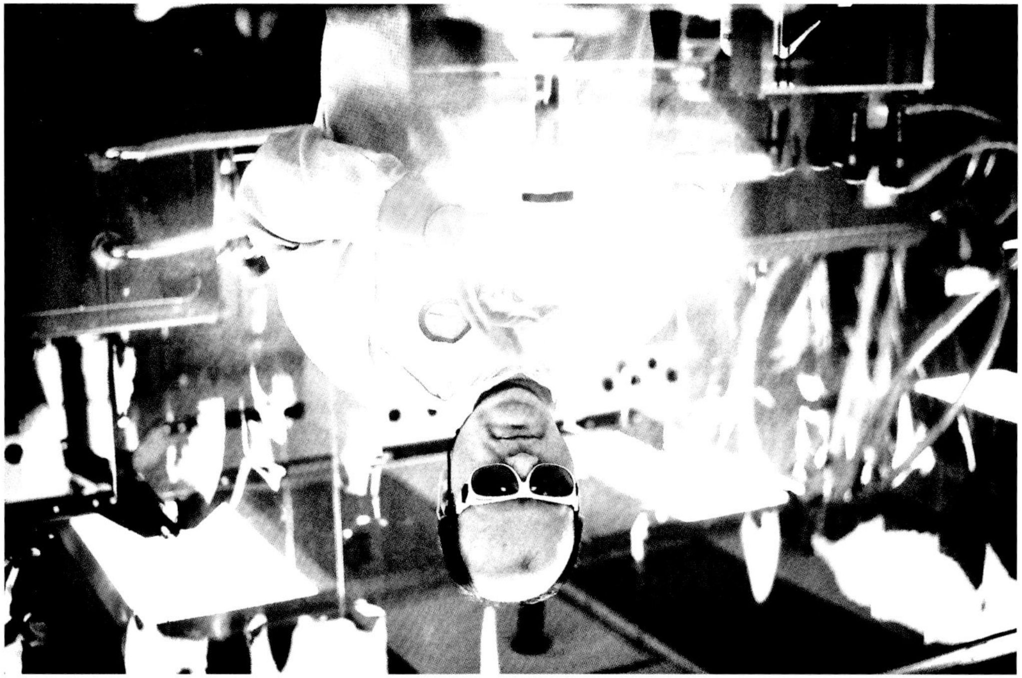

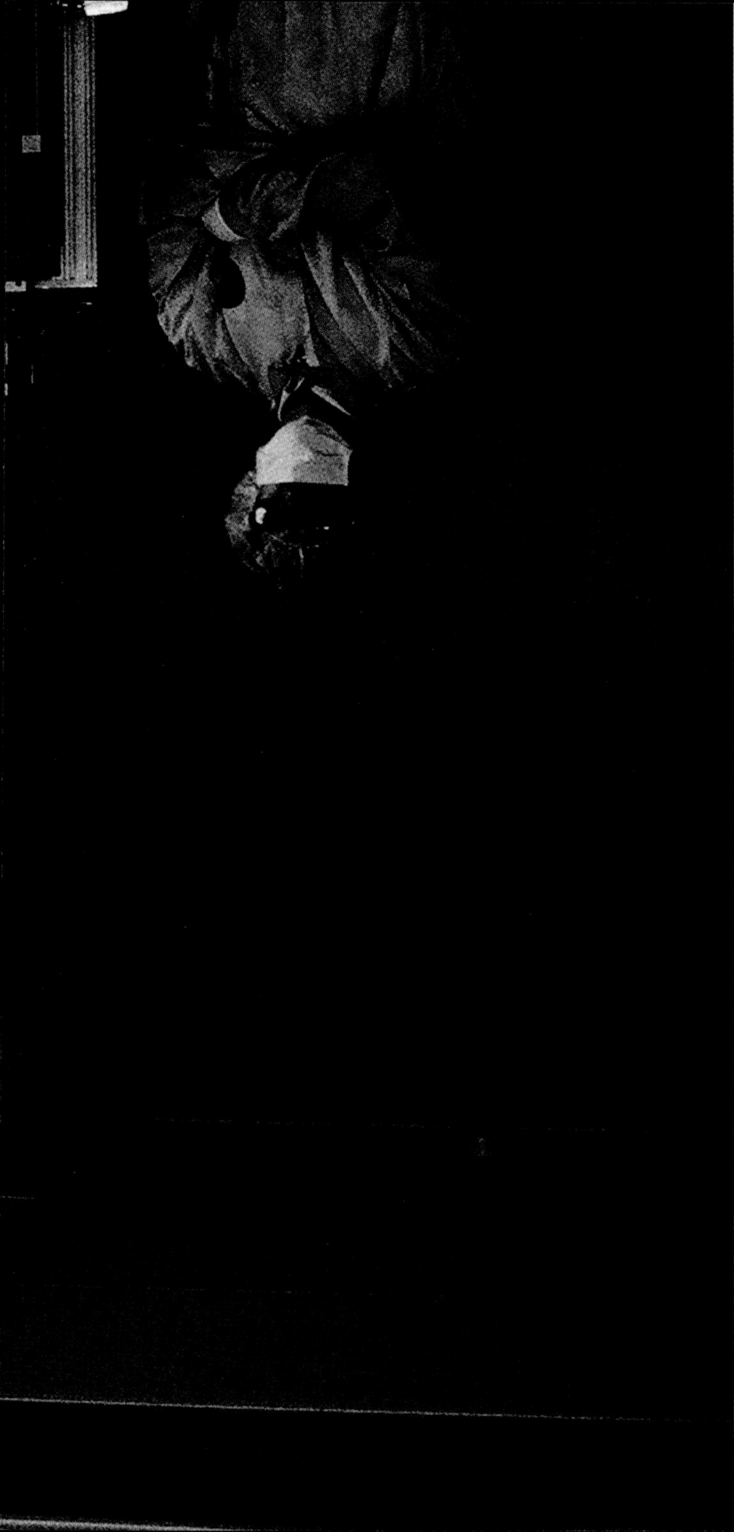

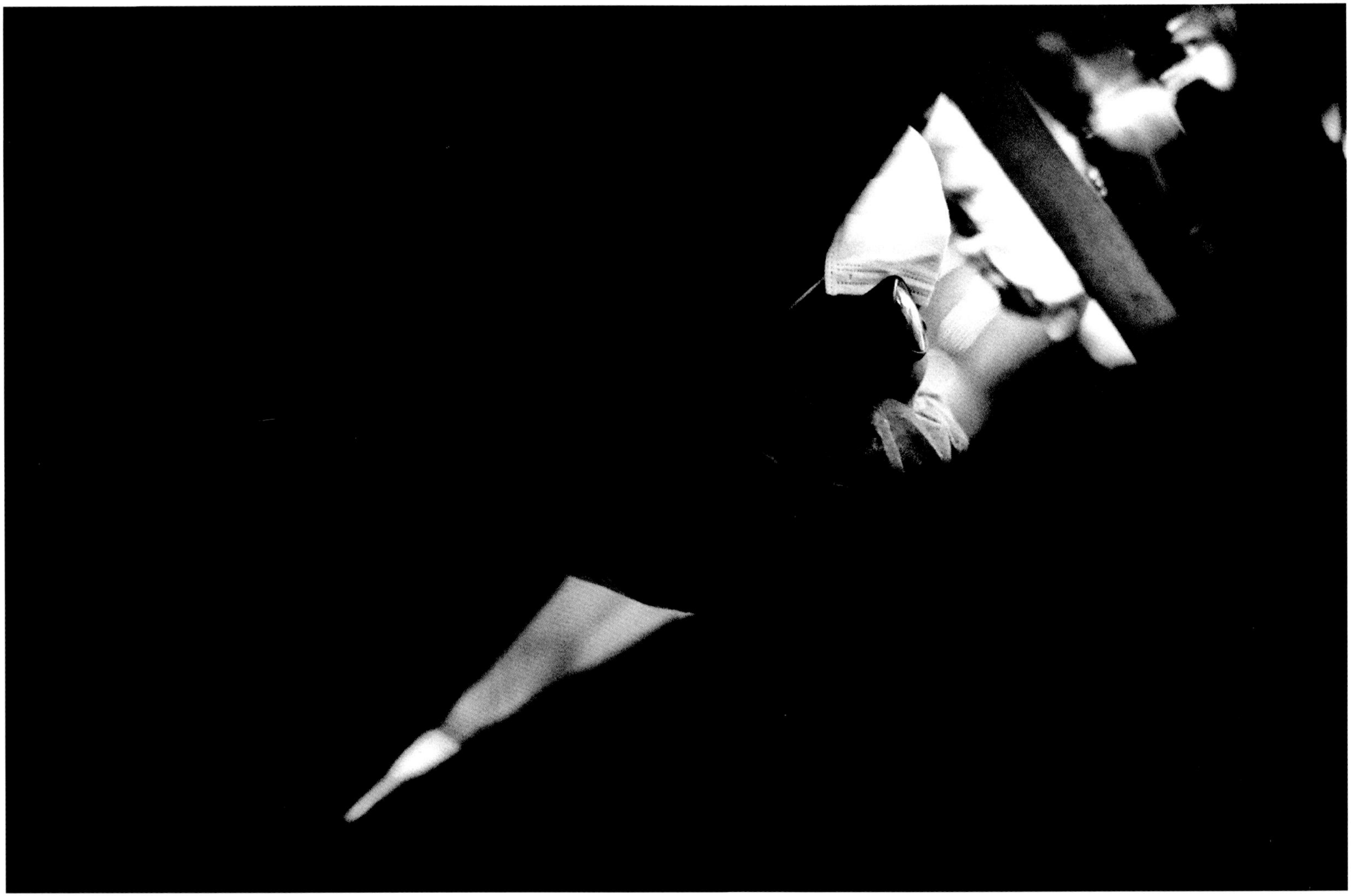

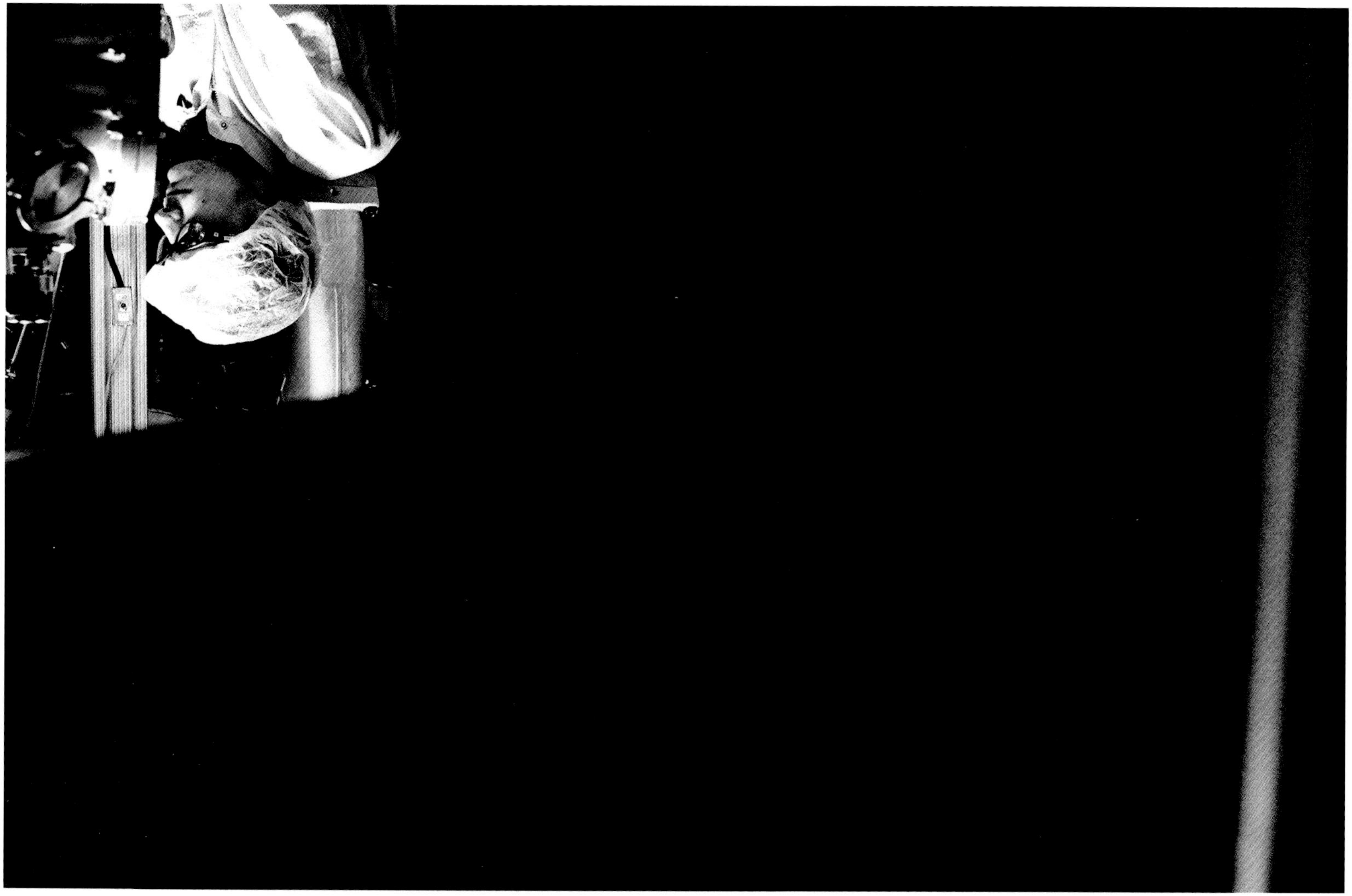

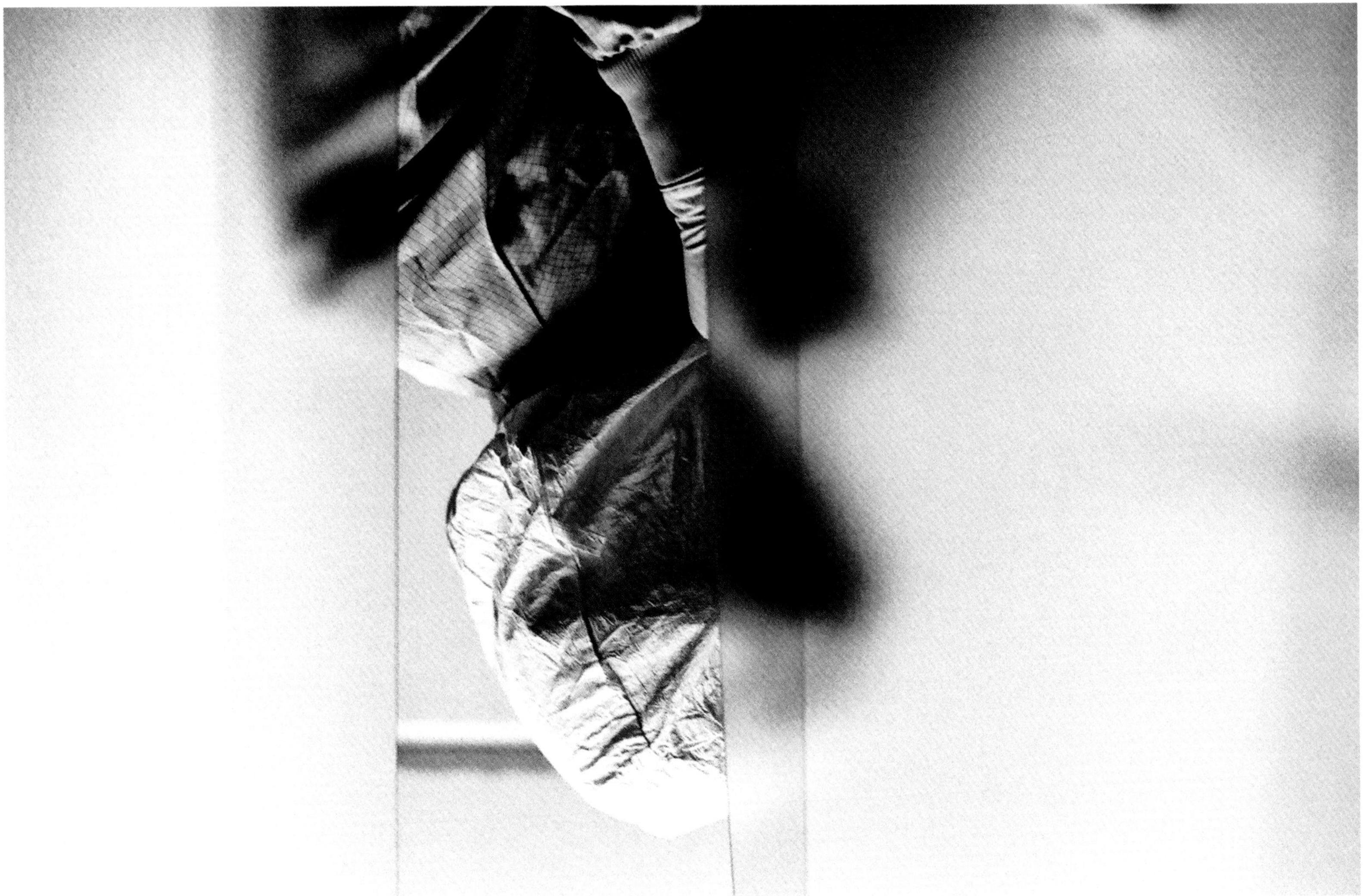

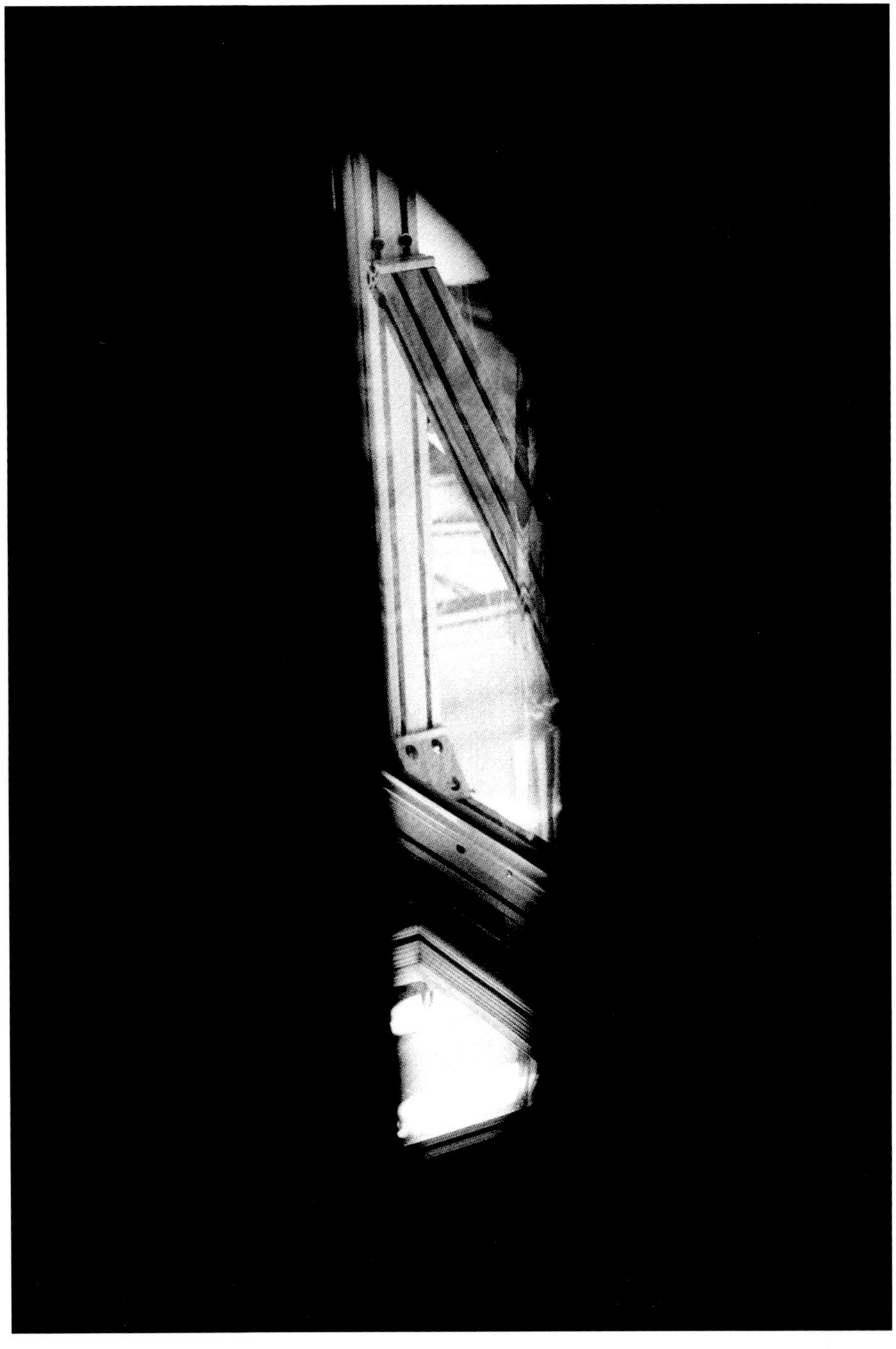

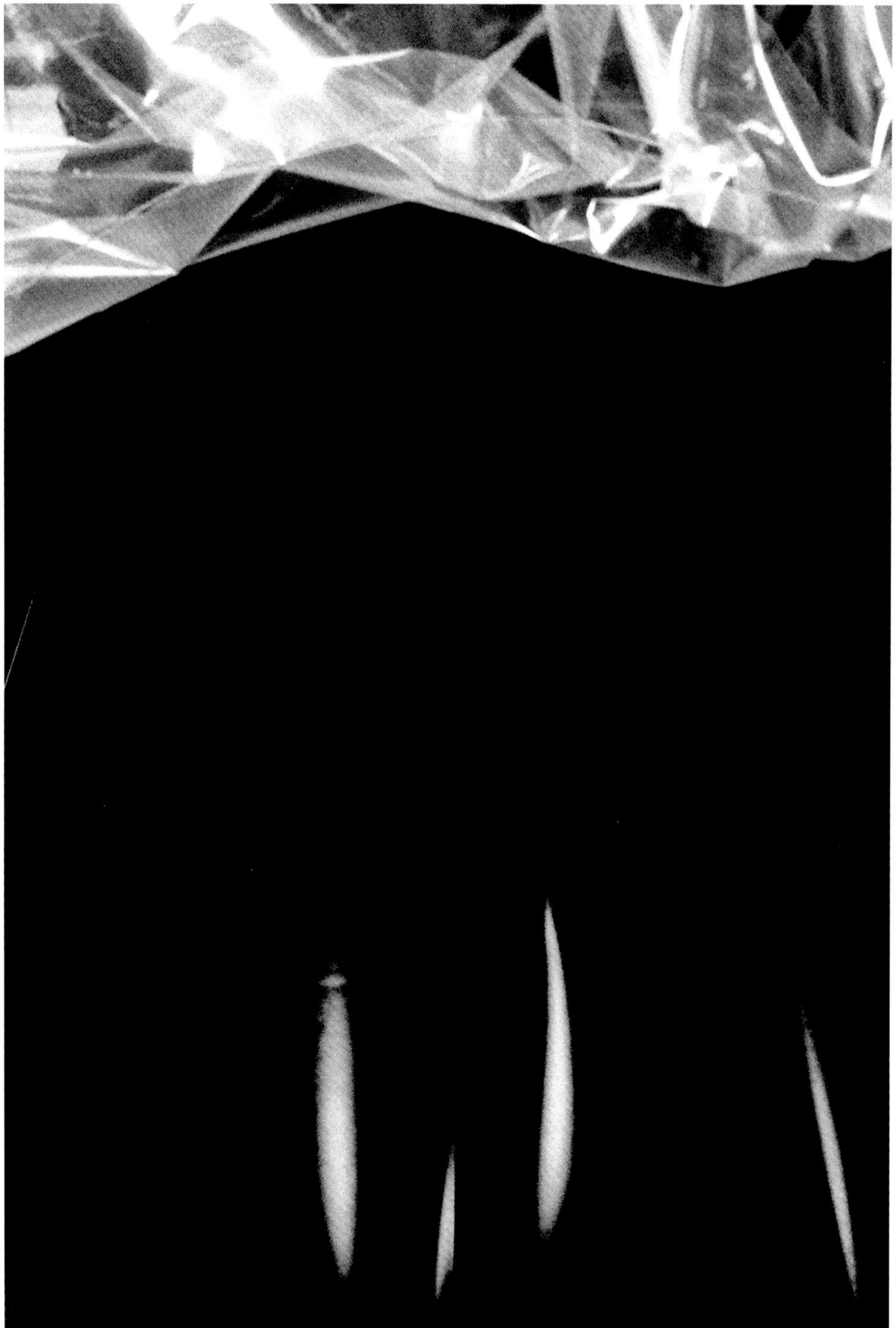

ZzzzzZZZttt!

The yottawatt laser beam drilled into the fleshy lump. The pulse was lasting much, much longer than usual, as if the biotech lump were impossibly sending signals up the beam to its source, jamming all the switches to On. First one, then two little stars bloomed within the shuddering bean. Though scorched and smoking, it held its shape. Jack still hadn't stopped singing. Adrian and Carla were backing away.

Fueled by the yottawatt beam and by the two pocket stars, the Yulia lump grew larger, taking form, extending arms, legs, and head, channeling energy like a babe at breast.

The thumping of the hidden vacuum pump had risen to a wild tattoo, and now came an explosion. The laser beam winked out. Somewhere in the lab an alarm horn was hooting. Perversely, idiotically, a set of ceiling sprinklers kicked on, raining down upon the scene. Jagged sparks, swirls of smoke, shattering glassware. The remaining lights cut out. Footsteps rushed to the lab door—Adrian and Carla escaping.

In the soft dark, wearing his laser shades, Jack could still see a little bit. Yulia was sitting up. Reborn. Smiling at him.

And now she opened her eyes.

Rudy Rucker received his bachelor's degree in mathematics from Swarthmore College and earned his PhD in mathematics from Rutgers University. In the course of a three-decade academic career, he taught at SUNY Geneseo, Ruprecht Karl University of Heidelberg, Randolph-Macon Woman's College, and San Jose State University. Rucker is regarded as a contemporary master of science fiction and one of the founders of the cyberpunk literary movement. He has received the Philip K. Dick Award twice. His 37 published books include the novels of The Ware Tetralogy and nonfiction books on the fourth dimension, infinity, and computation.

"Jack's already been in this lab before," said Adrian.

"And Carla's from a genomics research group," put in Jack. "She's bringing an add-on for Adrian's run." Adrian let the unexpected claim pass.

"OK, fine," said the feckless Cruz. "But I want you folks out of there in 10 minutes. Before the lab's next autoscan." He backed off and took a running jump across the hall. "Yes! I might even be on the Bolshoi level by then."

The laser lab was deserted, a bit sinister, with sagging cables, panels with jiggly readouts, workbenches like sacrificial plinths. The place was dimly lit, with stark pools of brightness in certain spots. Filtered through the laser shades, the potentially hazardous light came through in sour greens and tender mauves, in meaty reds and shinbone whites.

A vacuum pump was thumping, with a wheezing sound. "Chirped pulse amplification," said Adrian. "Like an accordion. Working the light up to the yottawatt level, back and forth, strong enough to zap protons to the petavolt scale. Enough to spark a pocket star. My setup is over here." They proceeded down the aisle, first Adrian, then Carla, then Jack.

Jack noticed an intense glow of infrared body heat coming off of Carla. She was scared. More than scared—terrified. Jack formed a sudden conviction that she was planning to sabotage tonight's run. Lurching into her from behind, he seized her wrist and pried the precious, crystalline case from her hand.

"Jack! *I'm* the expert on mitochondria."

"You killed Yulia, Carla. I have to say it. It was your fault. You did it on purpose. To get your hands on me." There. Laying it out at last.

Carla's voice rose by two octaves. "You are so *crazy*! I don't even *like* you anymore! Adrian! We need to get out of here!"

"Cruz said we have 10 minutes," said Adrian, not really understanding. "Be quiet and pay attention, you two. My target is right here on this little platform, a piece of foil. See Jack's mirror maze next to it? The laser pulse is going to make a pocket star. And then a magnetohydrodynamic vortex pulls it into the maze. Keep back. The pulse is coming in 10 seconds."

Jack shouldered past them, holding out his faceted box. He flipped a dark pink object onto the workbench—it was a fabslime-woven matrix for Yulia's mitochondria. Jack was singing, his voice liquid and weird. The magic bean was twitching like a pet.

"The champion pecan," she said after a bit, holding up a final nut. "He cracked all his friends. But can you even see? It's so dark tonight."

"We need our laser shades," said Adrian, pulling out two pairs of sunglasses. He handed one pair to Carla.

"You'll like these," Jack told Carla. "I've got my own pair in my car. I helped slimefab them for the lab."

They were standing by Jack's car now, an old-school convertible with its top down, a massive construct of Detroit steel. Cars like this were generally illegal to drive, but Jack had a historical preservation permit for his. He drew his pair of laser shades from the glove compartment, and now the three companions were standing there, goggling at each other, goofing on the scene. Although the laser shades had dark lenses, they had infrared laser crystals set into the rims of their frames.

"Ghostly," said Carla.

The crystals vibrate," said Adrian. "Scanning across the things you want to see. Scanning them with infrared, you understand, and the rays bounce back to your special lenses. So you're seeing a moiré image contour map. With pseudocolors based on temperatures. You look like a singer in a yowly music band, Carla."

She did a little dance in the street there, brandishing her faceted box and her champion pecan. Jack was in the driver's seat, ready to go. But now, as often happened, the car failed to start.

"I've been working on a fix," said Jack. He twisted around, rooted though the debris on his backseat, drew forth a crusty glob of fabslime the size of a coconut, and warbled an open-sesame command. Obligingly, the hairy orb split in two, revealing a glittering carburetor part, quite unobtainable on the commercial market. Jack flipped up the car's flappy old hood and installed the piece. Accustomed to this routine, Carla worked the starter until the car let out a dinosaur roar.

They cruised through the warm, dark Austin night, the three of them on the car's wide front seat, Carla in the middle, the air beating, pecans crackling beneath the wheels, the passing scenery like cartoons seen through their laser shades.

Adrian had to pass all kinds of thumbprint and eyeball scanner routines to get them down the elevator and as far as the actual entrance to the yottawatt laser lab. And then it became a matter of jollying their way past the gatekeeper, Cruz Sordo, who was somewhat distracted by a holographic ballet-dancing game.

"I've nailed my arabesques and fouettés," said Cruz, rocking back and forth. "I need three perfect grand jetés to reach the next level—which is the virtual Bolshoi. Your two guests are cleared, Adrian?"

"Maybe," said Adrian, not all that interested. "What's the second secret?"

Silently Carla mimed a juicy kiss.

"Carla is a postdoc in the mitochondrial genomics group at the university," said Jack, before Adrian could properly respond. "Specializing in the Golgi apparatus. She was working with Yulia right up to the end. She even found the fix to neutralize the prion that killed Yulia. Saved the others in the lab. They called her a hero."

"Soft, wet science," Carla told Adrian, her voice a tiger's purr. "Not like those yottawatt laser-beam swords you boys play with. Not like your pocket-pool hydrogen bombs. Genomics is the only femtotech that matters. A cornucopia from the living mother of life."

"The living mother of life, huh?" said Adrian with a crooked grin. "Does that have anything to do with your second secret?"

"Everything," said Carla. "Dim things *matter*, Adrian. Not just bright things." She was hefting the dark little container in her hand. It had sides like pentagons. "Take us into the laser lab, and this little stash box opens like a clam. You'll be flabbergasted."

"Say yes," Jack urged Adrian, his voice very intense. "You know I've been putting in an extra effort for you. And you've only let me into the lab that one time when you hired me. I need feedback if I'm going to keep working for you. Don't worry about Carla. I know all about her." A touch of ice in his voice.

"Carla is single?" asked Adrian, rudely direct. "Not your girlfriend?"

"Jack's the grieving widower," said Carla. "I'm the perky, yearning, starstruck ingenue, rejected once too often. And you can be Prince Charming, Adrian. If you don't act like a jerk. And if you're not too chicken to let your friends see what's in your lab. And if you really *do* have your pocket stars working. Which I'm starting to doubt." She paused for effect. "Maybe we should leave, Jack. I don't think I like this man." Carla rose to her feet, enjoying her power. She took two steps towards the door. Glanced back over her shoulder.

"Wait!" said Adrian, right on cue. He threw money on the table for the beers and followed Jack and Carla outside.

"I can drive," said Jack. "I've got my whale. I parked it down a side street."

"Great," said Adrian. "I came by bus."

It was a warm November night. The pecan trees were dropping nuts. Carla scooped up a handful, squeezing them together in pairs, eating the ones that gave way.

Carla giggled and nodded. “Kind of rank,” she said. “All burbly and wet. But maybe a little magical, too.”

Truth be told, Carla had once had a crush on Jack. She’d been Yulia’s research assistant, and with Yulia out of the picture, Carla had half expected to take her place. But nothing was happening along those lines, and Jack was getting ever stranger. Carla was about done with Jack. As a farewell, she’d let him rope her into helping him with this insane last-ditch scheme he was running tonight. Not that Carla even remotely expected it to work. Because if it *did*...but never mind that.

“I enjoy my work,” said Jack evenly. “How’s *your* project going, Adrian? Got those pocket stars happening yet?”

“Pocket stars?” said Carla, playing dumb. As if Jack hadn’t been steadily talking about this stuff for the last month. “What a beautiful name. Did you coin it, Adrian?”

Adrian would have liked to say yes, but he couldn’t. “This guy,” he said, jerking his thumb at Jack. “Good with words. I was going to call them femtoscale fusion reactors. You’ll use them like batteries, see. The technology of batteries is a millstone, a bottleneck, hopelessly stalled. Pocket stars will disrupt the paradigms.”

“What about hard radiation?” asked Carla.

“Not a showstopper,” said Adrian. “That’s the part I’m working on, matter of fact. Mirror mazes around our little suns. Phase-shift cancellations. Troughs and crests. Optical wizardry. That’s where Jack’s components come in.”

“How’s the latest upgrade working out?” asked Jack in a studiously neutral tone.

“Spectacular!” said Adrian. “We’re past the point of inflection, guy. Up onto the gigabucks slope of the growth curve. One more round of funding and my group can productize.” He lowered his voice. “The latest prototypes—they shed megawatts like dogs losing hair. I even sold some power to the lab. In the right matrix, one of these pocket stars could last indefinitely.”

“Can I see one?” asked Carla. “Pretty please.”

“Well, I wouldn’t be authorized to take you into the lab,” said Adrian. “It’s class-seven secure.”

“Oh, it’s Saturday night,” said Carla. “Nobody’s gonna be there. And I’ll show you one of *my* secrets if you’ll play.” She smiled, working her charm. “*Two* secrets, maybe.” She drew a little box from her purse, all angles, darkly gleaming, cupping it in the palm of her hand. “The first secret is that Jack slimefabbed something off a sample from my lab. Wouldn’t you love to know what it is?”

LASER SHADES
BY RUDY RUCKER

Adrian was entranced by Carla. She'd hooked him fast, and she was reeling him in—smiling with parted lips and nodding her head in rhythm to the cadences of his speech. Jack, off to one side, wasn't really listening to the words—no, he was reviewing tonight's plan. Step one: bump into Adrian. Step two: get into the laser lab. Step three...

This was a nice club, on Austin's merry Sixth Street, out towards the dark end of the spectrum. The Scales Fall. They featured yowly music here, one of Adrian's hobbies—he talked about the Scales Fall all the time, which was how Jack had known they'd find him here. Tonight a hairy guy was playing a "beam guitar," which was like a steel guitar but with sensitive light rays in place of the strings. The man wore his hair a hundred percent over his face, like a cartoon hermit, and the only skin you could see was the tip of his nose. A happy nose.

The beam guitar had a mellow, ethereal tone, sounding like one of those old-time gizmos—theremins. A woman was singing along, kind of a Russian steppes sound, her voice dank and husky, reminding Jack, as so many things did, of his dead wife, Yulia. Yesterday it had been six months. A prion infection from her lab. Horrible.

"Did you hear what Adrian said, Jack?" Carla was looking at him brightly. Humoring him.

"Uh, no," said Jack. "I'm lost in the music. A jellyfish." He made wiggly motions with his arms, managing to knock over one of their empty Shiner beer bottles. It bounced off the floor, unbreakable nanocrystal.

"Vintage slimefabber move," said Adrian, laughing at Jack. He was a tidy man with chiseled features.

"Slimefabbing is king," said Carla, sticking up for Jack. "Forget about brittle, thuddy machines. Jack cultures a wad of fabslime, he sings to it, and it makes what you need. Like the way a peach makes a pit."

"I know all about that," said Adrian. "Jack fabs components for my group at the yottawatt laser lab. I'm a plasma ultraoptics tech, right? Jack here's the only slimefabber in Austin who can make mirrored surfaces. You've known him for a while, huh, Carla? Have you ever heard him singing to his slime?"

Cofounders: Taj Forer and Michael Itkoff
Designer: Ursula Damm
Copy editor: Laura Fredrickson

ISBN 978-0-9897981-2-9

Printed in China

Daylight Books

E-mail: info@daylightbooks.org

Web: www.daylightbooks.org

ROBERT SHULTS

THE SUPERLATIVE LIGHT

Daylight